Introduction to
Smart eHealth
and eCare Technologies

T0314342

Devices, Circuits, and Systems

Series Editor
Krzysztof Iniewski
Emerging Technologies CMOS Inc.
Vancouver, British Columbia, Canada

PUBLISHED TITLES:

PUBLISHED TITLES:

PUBLISHED TITLES:

Tunable RF Components and Circuits: Applications in Mobile Handsets
Jeffrey L. Hilbert

VLSI: Circuits for Emerging Applications
Tomasz Wojcicki

Wireless Medical Systems and Algorithms: Design and Applications
Pietro Salvo and Miguel Hernandez-Silveira

Wireless Technologies: Circuits, Systems, and Devices
Krzysztof Iniewski

Wireless Transceiver Circuits: System Perspectives and Design Aspects
Woogeun Rhee

FORTHCOMING TITLES:

Diagnostic Devices with Microfluidics
Francesco Piraino and Šeila Selimović

Magnetic Sensors: Technologies and Applications
Laurent A. Francis and Kirill Poletkin

Nanoelectronics: Devices, Circuits, and Systems
Nikos Konofaos

Noise Coupling in System-on-Chip
Thomas Noulis

Radio Frequency Integrated Circuit Design
Sebastian Magierowski

Semiconductor Devices in Harsh Conditions
Kirsten Weide-Zaage and Malgorzata Chrzanowska-Jeske

X-Ray Diffraction Imaging: Technology and Applications
Joel Greenberg and Krzysztof Iniewski

Introduction to
Smart eHealth
and eCare Technologies

Edited by
Sari Merilampi • Andrew Sirkka

Managing Editor
Krzysztof Iniewski

CRC Press
Taylor & Francis Group
Boca Raton London New York

CRC Press is an imprint of the
Taylor & Francis Group, an **informa** business

CRC Press
Taylor & Francis Group
6000 Broken Sound Parkway NW, Suite 300
Boca Raton, FL 33487-2742

First issued in paperback 2020

© 2017 by Taylor & Francis Group, LLC
CRC Press is an imprint of Taylor & Francis Group, an Informa business

No claim to original U.S. Government works

ISBN-13: 978-1-4987-4565-9 (hbk)
ISBN-13: 978-0-367-65586-0 (pbk)

Visit the Taylor & Francis Web site at
http://www.taylorandfrancis.com

and the CRC Press Web site at
http://www.crcpress.com

Contents

SECTION I Opportunities and Barriers of Smart Technology in Care

SECTION II ICT-Based Platforms and Technology Examples

SECTION III   Case Studies and Field Trials

Preface

This book deals with very topical issues of the care sector by introducing possibilities of various technologies in health care. It also describes the need of these technologies in the care sector and discusses the requirements for the technology.

Aging population and a lack of resources in the care sector are serious concerns in many of the welfare states. The trend as well as the goal of today is to promote not only the older adults but all people with specialist needs to live as independently as possible instead of living in institutions. Owing to the lack of workers and other resources in the care sector, alternative methods and tools are needed to achieve this goal. ICT technologies provide enormous possibilities for these challenges. However, the unique user groups set very specific requirements of the technology. Specification of such user needs requires expertise of the care sector, service systems, as well as of the user group, which makes this kind of technology development work challenging.

The purpose of this book is to enhance communication between the two different fields of business: technology versus health and social care. Through this, the aim is to promote new innovations. This book provides information about various technologies to people working in the care sector. It also presents vital information to the technology developers about the real needs for technology as well as the requirements that need to be taken into consideration. This book will be useful for engineers developing well-being-related systems, software, or devices; for people working with people with specialist needs; for well-being and health service provides as well as for educators teaching related courses; and for upper-level undergraduate students and graduate students studying related topics.

This book is divided into "Front Matter" and three sections. The Front Matter gives an overview of the book and the authors. The first section, "Opportunities and Barriers of Smart Technology in Care," gives an overview of smart eCare and eHealth technology. The context as well as the needs, possibilities, and challenges of the technology are introduced. This section deals with the technology at a general level, and detailed information is provided in the next sections of the book. The second section, "ICT-Based Platforms and Technology Examples," discusses the technology first at the system level. Both hospital-related platforms and home care systems are introduced and discussed. After introducing platforms, the second section continues with chapters that focus on the device level. Examples on novel sensors are presented to illustrate ongoing developments and further evolving technical solutions. Common commercial sensors are also discussed as parts of the platforms. The third section, "Case Studies and Field Trials," combines issues discussed in the previous sections by introducing case studies and real-life pilots performed. Thus, the third section gives an insight into future trends and emerging solutions.

Sari Merilampi and Andrew Sirkka
Satakunta University of Applied Sciences

Series Editor

Krzysztof (Kris) Iniewski is managing R&D at Redlen Technologies Inc., a start-up company in Vancouver, Canada. Redlen's revolutionary production process for advanced semiconductor materials enables a new generation of more accurate, all-digital, radiation-based imaging solutions. Kris is also a founder of ET CMOS Inc. (http:// www.etcmos.com), an organization of high-tech events covering communications, microsystems, optoelectronics, and sensors. In his career, Dr. Iniewski held numerous faculty and management positions at University of Toronto (Toronto, Canada), University of Alberta (Edmonton, Canada), Simon Fraser University (SFU, Burnaby, Canada), and PMC-Sierra Inc (Vancouver, Canada). He has published more than 100 research papers in international journals and conferences. He holds 18 international patents granted in the United States, Canada, France, Germany, and Japan. He is a frequently invited speaker and has consulted for multiple organizations internationally. He has written and edited several books for CRC Press (Taylor & Francis Group), Cambridge University Press, IEEE Press, Wiley, McGraw-Hill, Artech House, and Springer. His personal goal is to contribute to healthy living and sustainability through innovative engineering solutions. In his leisurely time, Kris can be found hiking, sailing, skiing, or biking in beautiful British Columbia. He can be reached at kris.iniewski@gmail.com.

Editors

Sari Merilampi is the head of Well-Being Enhancing Technology Research Group at Satakunta University of Applied Sciences, Pori, Finland. She is also a founder member, shareholder, and board member of an electronics development company, Porel Ltd., in Finland. She earned a MSc degree in electrical engineering at Tampere University of Technology (TUT), Pori, in December 2006 and a DSc (Tech.) degree in June 2011. Sari's current research work includes new application possibilities of radio frequency identification (RFID), especially in the field of well being and sensing as well as well-being enhancing technology solutions in general. She has also worked with (mobile) games as a new tool for self-activation and rehabilitation. Sari has authored numerous publications in the field of passive ultrahigh frequency (UHF) RFID, new technological tools for the care sector, and materials and manufacturing methods of electronics. Her personal goal is to enable living to full capacity by means of person-centered technology. Sari can be reached at sari.merilampi@samk.fi.

Andrew Sirkka is a principal lecturer at Satakunta University of Applied Sciences (SAMK), Pori, Finland. Andrew's background is in health care (RSN, RNT) and adult education (Dr. Ed). He works as program coordinator of the Master Degree Programme on Welfare Technology (a joint program for engineers and health and social sector professionals), a member of the Course Management Team in EU Master Care & Technology International Master Degree Programme, and a member of the Well-Being Enhancing Technology Research Group at Satakunta University of Applied Sciences. Andrew has a long history of working as a visiting university lecturer, senior adviser in governmental agencies and international project activities, and a member of various international health scientific editorial and scientific boards. Andrew can be reached at andrew.sirkka@samk.fi.

Contributors

Indrek Ait
E-medicine Laboratory
Institute of Clinical Medicine,
    Technomedicum
Tallinn University of Technology
Tallinn, Estonia

Díez-Díaz Alvaro
Technology Transfer
PONS IP
Valencia, Spain

Beatriz Martinez-Lozano Aranaga
Health Ministry
Región de Murcia
Murcia, Spain

Serafin Arroyo
Marketing, Products and
    Projects—Innovation and
    Technology
Valencia, Spain

Lars T. Berger
Kenus Informatica
and
School of Engineering
University of Valencia
and
BreezeSolve
Parc Cientific
Valencia, Spain

Vicente Peñalver Camps
Kenus Informatica
Valencia, Spain

Maximo Cobos
Department of Computer Science
School of Engineering
University of Valencia
Valencia, Spain

Miguel Alborg Dominguez
IDI Eikon
Valencia, Spain

Enda Finn
Department of Computing Science and
    Mathematics
Dundalk Institute of Technology
Dundalk, Ireland

Nuno Garcia
Instituto de Telecomunicações
ALLab, Department of Informatics
University of Beira Interior
Covilhã, Portugal

Eric P.M. Hamers
Faculty of Bèta Sciences and
    Technology
Zuyd University of Applied Sciences
Heerlen, the Netherlands

Sirpa Jaakkola-Hesso
Faculty of Health and Welfare
Satakunta University of Applied
    Sciences
Pori, Finland

Andrea Kealy
Netwell CASALA
Dundalk Institute of Technology
Dundalk, Ireland

Antti Koivisto
Faculty of Technology
Satakunta University of Applied
    Sciences
Pori, Finland

Mirka Leino
Faculty of Technology
Satakunta University of Applied
    Sciences
Pori, Finland

John Loane
Netwell CASALA
Dundalk Institute of Technology
Dundalk, Ireland

Sari Merilampi
Faculty of Technology
Satakunta University of Applied
    Sciences
Pori, Finland

Margreet B. Michel-Verkerke
Researchgroup Technology,
    Health & Care
Social Work and Technology
Saxion University of Applied Sciences
Enschede, the Netherlands

Díez-Díaz Mónica
Veterinary and Experimental Sciences
    Faculty
Catholic University of Valencia
Valencia, Spain

Juan J. Perez-Solano
Department of Computer Science
School of Engineering
University of Valencia
Valencia, Spain

Nuno Pombo
Instituto de Telecomunicações
Department of Informatics
University of Beira Interior
Covilhã, Portugal

López-Moya J. Rafael
Patents Department
PONS IP
Madrid, Spain

Andrew Sirkka
Faculty of Health and Welfare
Satakunta University of Applied
    Sciences
Pori, Finland

Madis Tiik
E-medicine Laboratory
Institute of Clinical Medicine,
    Technomedicum
Tallinn University of Technology
Tallinn, Estonia

Riikka Tupala
Faculty of Health and Welfare
Satakunta University of Applied
    Sciences
Pori, Finland

Pauli Valo
Faculty of Technology
Satakunta University of Applied
    Sciences
Pori, Finland

Johanna Virkki
Department of Electronics and
    Communications Engineering
Tampere University of Technology
Tampere, Finland

Charles G. Willems
Faculty of Health Care Research Centre
    for Technology in Care
Zuyd University of Applied Sciences
Heerlen, the Netherlands

Section I

Opportunities and Barriers of
Smart Technology in Care

1 Smart eHealth and eCare Technology
What Is That?

Sari Merilampi, Andrew Sirkka,
Riikka Tupala, and Sirpa Jaakkola-Hesso

CONTENTS

The use of technology in health and social sector has increased lately due to its huge potential in solving challenges caused by the aging of the population, in helping the healthcare professionals' work, and especially in enabling living to full capacity despite possible personal limitations and disabilities. Nowadays, the technology can be implanted or wore, or may be part of our environment. Information and communication technology (ICT), electronics, and automation have already had a huge effect on various fields of business. The same technology that is targeted for industrial use has also many applications in health and well-being. The demographics and lack of resources in health and well-being industry are increasingly forcing us to find alternative solutions for more individualized care or social services. Smart technologies provide enormous potential for solving this challenge. However, technology industry, healthcare industry, and social services sector have not yet been accustomed to collaborate deeply that is needed to make the technology serve this purpose. Smart eHealth and eCare Technology is *no-mans-land* due to its multidisciplinary nature, and very few people have adequate expertise and experience from all fields concerned—technology, healthcare, and social sector. The lack of common understanding has led to wrong kinds of technology products that do not

3

provide expected user-friendly solutions needed in real-life situations and service delivery systems. Nevertheless, the time is now ripe to foster joint development between the two different fields of business in order to create innovative solutions to meet the real user needs with the help of the latest technology achievements. A good variety of technological innovations are already available for various purposes including numerous assistive technologies and barrier free design supporting autonomous living, novel sensor technology, mobile applications for activation and self-rehabilitation purposes, electronic patient records system for improving care and communication between various parties in the healthcare delivery systems, reliable medical technology for improving treatment and analytics, and even mobile phone-based testing applications for making laboratory analyses easier and more cost effective (Häyrinen 2008, Karlen 2014, Kiili 2010, Koivisto 2011, 2013, 2015, Merilampi 2014, Sirkka 2012, Wichert 2008).

This book enhances communication between technology, healthcare, and social sectors, and tends to promote new innovations. The purpose of this publication is to discuss various technologies and share experience and knowledge regarding the needs and requirements for developing technology products successfully. This discussion starts with terminology clarification and introduction to various types of assistive and enabling technologies.

This book defines smart eHealth and eCare technology as the use of technology in promoting health, well-being, and quality of life, and thereby assisting people with impaired or declining function (physical, psychological, cognitive, or social). The term *user* may refer to an individual with declined functioning, a professional, or a family caregiver. In most cases, the benefit of using technology is typically real-time data about issues related to the patients' status of health and well-being, or the data related to therapeutics or other caregiver actions.

This chapter tends to provide a general overview of the available assistive and enabling technologies. The subject is very wide, and it is hard to define which technologies can be considered enabling ones and which not. According to Löfqvist (2005), assistive technologies could be divided into two categories: (1) *low technology* referring to more traditional and mechanical assistive technology (such as grab bars) and (2) *high technology* referring to devices and equipment based on more advanced technologies. This book will focus on the high technology in which ICT, electronics, and automation technologies play an important role. Care technologies are further discussed as seven partially overlapping categories:

1. Assistive (communication and information) technologies
2. Safety and social technologies
3. Health technologies
4. Self-activation and personal development technologies
5. Design-for-all and ambient assisted living (AAL) technologies
6. Gerontechnologies
7. Hospital technologies and electronic health records systems

Each category of care technology is briefly discussed in the following sections. Examples of related devices, equipment, and applications are introduced, and common terminology used in the literature is explained. This book contains more detailed chapters about the different categories of technologies, and the related chapters are referred in the text (Löfqvist 2005, Suhonen 2007).

1.1 ASSISTIVE TECHNOLOGY

Assistive technology is a very wide category among enabling technologies. It includes tools, equipment, and devices that are used to facilitate the operation and survival of a person with limited capabilities. In general, assistive technologies assist an individual to perform daily tasks or prevents injuries. These technologies compensate sensory, physical, and cognitive impairments, and promote safety for vulnerable individuals in terms of detecting and reporting health hazards. Assistive technology consists of devices that aid individuals in communication, movement, controlling the surroundings, performing daily activities, and in overall life management. Noncomputer-based assistive technologies include items such as wheelchairs, grab bars, Braille, and general solutions that make home environment more accessible. Technically more advanced equipment includes environmental management devices and videophone network. Assistive technologies also include a variety of equipment used in medical care, disease prevention, diagnostics, and rehabilitation (Huuhtanen 2012, Lindeman 2009, Salminen 2010, http://www.papunet.net, http://www.thl.fi/apuvälineet, http://www.tikoteekki.fi).

1.1.1 INFORMATION AND COMMUNICATION TECHNOLOGY

Information and communication technology as a subcategory of assistive technology includes technological devices that assist in reading, writing, speaking, and using a computer or a phone. *Information technology* includes special assistive equipment (software and hardware) enabling computer and mobile phone use, such as special key boards and mouse. Button mouse is an alternative to a standard computer mouse. The buttons of the mouse are easier to observe and push, and they can be customized according to the user's needs and capabilities. In a head mouse, the head movements are followed with the help of a camera. Screen readers (software) can be used to read aloud text on the screen. Screen magnification software is another example of assistive computer software. Small portable electronic magnifiers or video magnifiers that can be used to magnify printed text or pictures are also available. In addition, assistive devices include alerting devices (such as blinking light) that can be connected for example to a telephone to indicate ringing (and for many other actions) (Huuhtanen 2012, Lindeman 2009, Suhonen 2007).

Communication technology in terms of augmentative/alternative communication can be used to assist communication and social interaction in the case of temporary or permanent loss of speech ability. Alternative communication is used when a person is not able to speak, and augmentative communication is used when the speech is unclear or incomplete. Visual communication with graphics and signs are commonly used (pictures, photographs, written words, Bliss symbols, picture communication symbols, pictograms etc.). Technology to support augmented or alternative communication is apparent in devices that play prerecorded voice messages and in hand-held devices that show a written message on the display and talk the text aloud. The text can be produced from symbols/signs as well. The communication panel can be controlled with fingertips, but alternative methods such as eye tracking and a forehead stick are also available. Computer-assisted communication in this context refers to a speech synthesizer in which a combination of computer/mobile devices and communication programs are most commonly used. Mobile apps are also available to assist communication. A smartphone or a similar device may work as a communication tool in face-to-face situations, but the symbols in the app can also be used to generate messages. The symbol-generated messages can be received as text messages on other phones. In addition, to communicate with various communication devices, a traditional communication folder is always needed. An example of a traditional communication folder system in Finland is *TAIKE- board* that works together with related *Speaking Dynamically communication* software (Huuhtanen 2012, Lindeman 2009, Salminen 2010, http://www.papunet.net, http://www.thl.fi/apuvälineet, http://www.tikoteekki.fi).

Various sensors and ICT-related equipment can be used to develop assistive technology or as part of assistive devices. Some examples of novel sensor technology suitable for assistive devices are presented in Chapters 6–9. Generally speaking, while developing software, the special user groups (like older adults with naturally declining functions) should also be taken into account (e.g., Internet pages, user interfaces, and games). Design-for-all in terms of producing more accessible and usable software does not require huge efforts from developers; rather, it is only a mindset. Examples of mobile games and applications for special user groups are presented in Chapter 13.

1.2 SAFETY AND SOCIAL TECHNOLOGIES

In common with safety and social technologies is the aim of increasing security and independence of a person who typically has some limitations in functioning. The systems assist the person to be active by increasing the level of accident prevention and thereby the sense of security. In cases of an accident, the system assists to get the help fast. The technology can be implemented in a person's home or in an assisted living environment.

Safety technology includes different security systems easily interlinked with other assistance, care, and surveillance services. Safety bracelets, safety phones, and additional integrated sensors are part of these systems. Traditional safety bracelets are passive in nature and require user to push a button to make an alert and call for help from a relative or a care provider. *Smart bracelets* measure the users' activity and movements triggering the alert when any abnormality is observed in the registered data. The most advanced solutions offer possibilities to monitor activity-sleep levels and GPS tracking. Integration of different sensors such as smart floor, carpet, contact switches (doors, windows), bed occupancy sensors, and motion sensors enables a provision of extensive and complete safety systems. In addition to various sensors, the systems can be equipped with actuators (e.g., safety switches) and health technology (e.g., biomedical monitors). The devices operate in a network connected to a remote center for data collection and processing. The remote center monitors the situation and initiates assistance procedures if required. The technology can be extended to wearable and implantable devices to monitor people 24 hours a day both inside and outside the house. The term *smart home* is typically used in this kind of safety-system-related context, although smart homes typically also include nonsafety-related equipment (such as building automation for energy management and indoor environment control) (Chan et al. 2009, Suhonen 2007).

Social technology is quickly emerging as part of home care systems. In addition to safety, social technology provides social interaction and virtual services such as active sessions remotely with caregivers and other service providers. These kinds of systems are important to avoid social isolation. They are also very practical solutions to provide services in rural areas. The aim of the technology is to assist in independent living by modern ICT through remote guidance and virtual services (see also health technology). Social networking technologies enable the creation of social networks and focus on building communities of interest that help users communicate, organize, and share with other users and with their care providers. Health TV is an example of social technology. It provides visual and speech connection (via broadband connection) from an easy user interface of the TV (Lindeman 2009, Suhonen 2007).

Although the aim of safety and social technology systems is to improve the quality of life, there are challenges are as well. These challenges include false alarms and ethical issues related to the privacy of the user as well as data-handling properties, responsibilities, and rights, and also practical issues related to big data handling.

Social technology and safety are discussed further in Chapters 6 and 14. Sensor technologies and their possibilities are discussed in detail in Chapters 6–9.

1.3 HEALTH TECHNOLOGY

Health technology includes self-monitoring- and self-care-related systems and equipment. These systems include monitoring and measurement devices (such as blood pressure monitors and other equipment typically used by healthcare professionals), ICT-related devices (mobile devices and computers), data connection (the Internet and mobile), and remote communication with medical professionals.

The term *remote patient monitoring* is typically used in this context. A wide variety of technologies are designed to manage and monitor a range of health conditions. Point-of-care (e.g., home) monitoring devices, such as weight scales, glucometers, and blood pressure monitors, may stand alone to collect and report health data, or they may become part of a fully integrated health data collection, analysis, and reporting system that communicates to multiple nodes of the health system and provides alerts when health conditions decline. Internet-based services include eHealth portals that offer health services, self-care services as well as healthcare and well-being products and information via the Internet. For example, a user can get consultation from a doctor via these portals and access different health-related tests. A simple example of a relatively common health technology is an Internet-based self-care assisting system for monitoring blood sugar levels as part of diabetes self-care. In the system, user provides sugar levels to a server where they are available to healthcare professionals for monitoring. Similar systems are also provided for remote blood pressure monitoring as well as obesity, allergy, and asthma treatment (Lindeman 2009, Suhonen 2007).

Chapter 10 provides an insight into the dilemmas and issues related to self-care systems and patient information sharing. Additional discussion can also be found in Chapter 5.

1.4   SELF-ACTIVATION AND PERSONAL DEVELOPMENT TECHNOLOGY

A fast developing and growing area of technology is the development of software, equipment, and systems used to support, monitor, and analyze personal development and performance. Many of these devices are first targeted for athletes for performance monitoring during a workout, but later they become very common for *ordinary people*. Typical *sports technology* includes pedometers, activity bracelets, and mobile apps that monitor activity levels and also remind the user to be active and cheerful. Heart rate monitoring is also very common. First separate heart rate belts and watches and later smart watches with integrated heart rate sensors are commonly used for investigating heart rates during a workout such as running or cycling. Smart sports watches typically provide other data as well. GPS allows the user to see the distance as well as speed. Some smart watches also have interface for making programs that utilize the sensor data for various purposes such as for sleep monitoring, epileptic seizure detection, activation, and stress controlling. Sensors such as accelerometer, gyroscope, magnetometer, pressure sensor, heart rate monitor, activity sensor, GPS, temperature sensor, and even sweat sensors are already in use or in the development phase. In the future, even more sophisticated sensors will be integrated into these wearable equipment (such as sweat analysis), making them even more useful for health and well-being. Many kinds of different sensors are also being developed for more precise purposes such as trajectory monitoring and monitoring of bodily functions. Although the technology may not originally be targeted for healthcare purposes, it provides huge possibilities after further development for rehabilitation and self-care. This equipment has the potential for encouraging people to be more active by 1) providing interesting data about the personal progress or 2) by increasing the feeling of safely because the bodily functions can be monitored and healthy stress level thus be maintained (Empatica E4 wristband 2015, Heikenfeld, 2014, Phillips 2014).

Gamification is also one of the recent trends. It means to make activities more motivating, playable *game like*. To give an example, gamification in health and well-being context may refer to making repetitive rehabilitation/exercising more motivating by providing some sort of meaning for the action such as a progress in a game. Health and well-being games may relate to physical fitness/rehabilitation as well as cognitive fitness/rehabilitation, and they typically contain some sort of tracking and assessment component. The game analytics offer tools and possibilities for diagnosing purposes and progress monitoring (Kiili 2010, Koivisto 2011, 2013, 2015, Lindeman 2009, Merilampi 2014, Sirkka 2012).

The term *exergame* is typically used when referring to games in which progress depends on physical exercise. Commercial game consoles already provide wireless game controller that may be used to track body movements and control various commercial exergames. Cognitive fitness and assessment technologies include thinking games and cognitive challenge regimens. Like physical fitness, the premise of cognitive fitness is that cognitive health can be maintained or improved if individuals exercise their brain. The emphasis may be, for example, to prevent or delay Alzheimer's and related dementias. Many cognitive fitness technologies are computer or the Internet-based, and include assessment and tracking component. The health and wellbeing games can also be educational. Games also provide solutions for recreation, and by right kind of design, games can also provide various possibilities for special user groups. Mobile health games are discussed in Chapters 3 and 13 (Kiili 2010, Koivisto 2011, 2013, 2015, Lindeman 2009, Merilampi 2014, Sirkka 2012).

1.5 DESIGN-FOR-ALL AND AMBIENT ASSISTED LIVING TECHNOLOGY

In accordance with the *design-for-all* philosophy, everybody should be capable of participating in our society, with equal opportunities regardless of personal characteristics such as age, gender, ability to function, and cultural background (Design for All Foundation 2015, Kemppainen 2008, 16). When talking about equal participation, terms such as accessibility, usability, and universal design are also used. The meaning of all these terms is close to each other, and usually the context defines the usage of these terms. Terms *universal design* and *design-for-all* both mean that products, services, and environment should suit everyone or at least as many as possible, whereas built environments can be accessible and devices usable. Also term *availability* is commonly used when talking about the availability of information, services, products, and so on. However, it is important to remember that even though information is available, it is not automatically accessible.

It has been shown that for the 10% of the population accessibility is essential, for 40% necessary, and for 100% it is comfortable (Design for All Foundation 2015). Sometimes, solutions that are implemented for special features bring benefits to other users and increase overall usability and broaden the market. For example, the control of volume amplification in telephones was originally developed for people with hearing problems but was also found useful for anyone using a telephone (Mellors 2004, 15).

To be able to provide equality and equal participation, there must be products, services, and environments that are suitable for everybody. The ideal situation is when a single solution is suitable for all potential users and does not need adaptation depending on the user. That calls for good designing and planning. Sliding doors are good example of well-designed product that represents all the design-for-all criteria. The product respects the diversity of users, so that nobody feels marginalized and everybody is able to access it. The product is also safe to use and does not cause risks for health. Functional and comprehensible use is fulfilled as sliding doors always work in the same way and in most cases work automatically. The product is also sustainable, affordable, and appealing (Design for All Foundation 2015).

Isolated and special accessibility solutions should be avoided in designing and planning. These kinds of solutions do not directly improve design-for-all thinking and acceptance of diversity. Furthermore, it is not cost-effective and beneficial. Nevertheless, solutions cannot always be designed to be suitable for all (Mellors 2004, 15). For example, products such as prostheses need to be customized individually. Also, the wider the range of different sizes and models of shoes available, the better. If a product, service, or environment could not be made for all or it is not practical to be designed for all, it could be either adjustable, individual (range of products), compatible with commonly used accessories, or customized product or service. In some cases, it is also acceptable that solutions are compensated with good services or alternative solutions to the mainly used offering (usually solution made afterward) (Design for All Foundation 2015).

Design-for-all philosophy is not only of benefit to the end user, it can also offer benefits to business (Mellors 2004, 15). If possible, involvement of both the end users and the experts to all stages of the design process is recommended. Solutions that are designed according to design-for-all philosophy and in cooperation with the end users are usually effective and do not need further adaptations. These kinds of solutions and involvement also increase customer satisfaction (Aragall et al. 2013, 14).

Ambient assisted living (AAL) technology is related to design-for-all as it aims at living to full capacity despite limitations in the ability to function. AAL is one of widest categories presented, and it covers most of the technology described in this chapter except the hospital technology. However, typically AAL refers to elderly care.

In Pieper (2011), AAL is defines as follows:

> Ambient Assisted Living (AAL) comprises interoperable concepts, products, and services that combine new ICTs and social environments with the aim to improve and increase the quality of life for people in all stages of the lifecycle. AAL can at best be understood as age-based assistance systems for a healthy and independent life that cater to the different abilities of their users. It also outlines that AAL is primarily concerned with the individual in his or her immediate environment by offering user-friendly interfaces for all sorts of equipment at home and outside, taking into account the fact that many older people have impairments in vision, hearing, mobility, or dexterity. Thus, it implies not only challenges but also opportunities for the citizens, the social and healthcare systems as well as the industry and the European market. The roots of AAL are in traditional assisted technologies for people with disabilities, design-for-all approaches to accessibility, usability, and ultimately acceptability of interactive technologies, as well as in the emerging computing paradigm of Ambient Intelligence, which offers new possibility of providing intelligent, unobtrusive, and ubiquitous forms of assistance to older people and to citizens in general.

To give some more examples, AAL covers smart home technologies discussed earlier in this chapter and many assistive equipment and devices such as robots and medication optimization equipment. Interactive robots may cooperate with people through bidirectional communication and provide personal assistance with everyday activities such as help them prepare food, eat, and wash or remind them to take medicine. The term *medication optimization* refers to a wide variety of technologies designed to help manage medication information, dispensing, adherence, and tracking. Technologies range from the more complex, fully integrated devices that use information and communication technologies to inform and remind stakeholders at multiple decision and action points throughout the patient care process to the simpler, standalone devices with more limited functionality (Flandorfer 2012, Lindeman 2009, Wichert 2008).

It is also worth mentioning that nowadays the trend in modern households is to add intelligence and building/home automation such as remotely controllable or measurable devices and systems. The so-called Internet of things (IoT) in which smart devices communicate and collaborate (collect and exchange data) is very interesting in considering AAL, because they also have huge potential for assisted living. Adjustable and controllable lighting, temperature, moisture, air conditioning and ventilation, venetial blind, remotely measurable electricity consumption, remotely activating power sockets, and even household machines such as smart washing machines, stoves, refrigerators, televisions, WLAN cameras, audio equipment, and even smart self-watering flower pots are already available. Also smart and more secure locking systems and fire/smoke alarm systems are commercially available. The house key can be even implanted in a hand of a family member and access rights may be controlled with software. In addition, furniture and textiles are also becoming smarter and making the life easier.

AAL is discussed in Chapters 6 and 14. Sensor technologies and their possibilities are discussed in detail, for example, in Chapters 6–9.

1.6 GERONTECHNOLOGY

Because of the demographic situation of welfare states, the role of gerontechnology in well-being has been significant. Gerontechnology is a fairly general term. It includes technology from all the above discussed categories, and thus it is not discussed with examples but at more general level. The goal of gerontechnology is to develop age-friendly technology as well as help older adults to use existing technology. The ultimate goal of gerontechnology is to increase the older adults' quality of life meaning being socially active, healthy, and independent up to high age. Gerontechnology aims at preventing age-related problems and supporting older adults' strengths, compensating declining ability to function, supporting elderly care, and promoting related research (such as research and development to provide age-friendly product and service designs). Although gerontechnology is targeted for promoting well-being of older adults, the technology solutions are typically useful for other special user groups as well (Kuusi 2001, Suhonen 2007, Väyrynen and Kirvesoja 1998).

1.7 HOSPITAL TECHNOLOGY AND ELECTRONIC HEALTH RECORDS SYSTEMS

In this book, the term *hospital technology* does not refer to medical technology and devices such as surgery robots, EEG and EKG monitors, and other very special equipment used in treatment and diagnostics in hospitals. This book discusses smart eHealth and eCare technologies, and medical technology is outside the scope of it. However, hospital data and information systems such as electronic health records (EHR) are discussed in Chapter 5.

EHR is a digital version of the traditional patient paper chart. However, the idea of an EHR is to combine traditional patient information (medical and treatment information) into other information used in health and social care. The definition of EHR and the data content is difficult due to various practices used. According to a review (Häyrinen 2008), the concept of EHR comprised a wide range of information systems, from files compiled in single departments to longitudinal collections of patient data. EHRs are used in primary, secondary, and tertiary care. The data could be recorded in EHRs by healthcare professionals and secretarial staff (recorded data from dictation or manual notes). Some information is also recorded by patients themselves (validated by physicians). Several data components are documented in EHRs, such as daily charting, medication administration, physical assessment, admission nursing note, nursing care plan, referral, present complaint (e.g., symptoms), past medical history, life style, physical examination, diagnoses, tests, procedures, treatment, medication, discharge, history, diaries, problems, findings, and immunization (Häyrinen 2008, Suhonen 2007).

Hospital information systems and EHRs are essential considering various above mentioned enabling technology devices and home care type systems and the information they produce. This information should be supported by the service system in general (politically) as well as by the patient records systems (technically). Integrating data from smart home-type systems with data from electronic patient or health records is still in an early stage. Several projects are in an advanced conceptual phase, some of them exploring feasibility with the help of prototypes. General comprehensive solutions are hardly available. In addition, a global electronic patient record system would be a huge achievement, but, for example in Finland alone, mainly three different patient systems are used. The challenge to build compliant systems is huge because the systems nowadays are different in each country. Law and legislations in different countries also differ from each other, which has to be considered as well (Chan et al. 2009, Knaup 2014, Suhonen 2007).

ICT-based service platforms and patient records systems, services, service systems, and data handling are discussed in more detail in Chapter 5.

1.8 PERSON-CENTEREDNESS OF THE TECHNOLOGY

Despite the category the technology belongs to, the most important issue to take into account in eHealth and eCare technology is the person-centeredness. Technology has to meet the real-user needs and serve its purpose. The technology must be usable and accessible and in many cases as *invisible* as possible. Technology resistance will typically arise if these requirements are not met. The user needs may not be obvious but rather challenging. For example, if a user with dementia wears a health bracelet, it provides safety and information independent on the location of the user. The technology is also easy to implement because no construction work or clinical operations is needed. However, the device may be scary and look suspicious, and it is then removed. In this case, the user may benefit if the technology would be implanted, embedded into clothing, hidden as part of the apartment, or as part of ordinary and traditional equipment.

Additionally, the technology should not mean *extra work* for the professional staff or the individuals but in the contrary. Attention must be focused on implementing the technology: integrating it as part of the service system, offering easy procurement and distribution, deployment, and training as well as maintenance. *Design-for-Somebody* is a technology design and development philosophy defined by the well-being technology research group at Satakunta University of Applied Sciences (SAMK), meaning that not only special user groups' needs but also an individual's needs should be placed as the centerpiece in technology and service development processes. By going very deep into user needs, key features are found and high customer satisfaction is achieved. Although technology cannot be actually developed for each person separately, there are many similar persons to whom the same technology can serve. For example, the same technology can be modified according to a person, making it more individual, although 99% of it is similar to other users as well. For successful identification and interpretation of individual needs, the research, education, and working life experts from the fields of healthcare, well-being and technology, business, and so on. must be employed. Development processes must be conducted in close collaboration with beneficiary parties and organizations. Research and development of eHealth and eCare technologies requires constant and multidisciplinary knowledge updation and agility to response to changing demands and circumstances, making it a very challenging subject but also serving endless possibilities. Design-for-Somebody philosophy is discussed further in Chapter 3.

One of the challenges related to enabling technology is to measure the functionality and effectiveness (and revenue) of the technology. The research focus is shifting from pure technology development to usability and user experience studies, which is a good start as nonaccessible and nonusable technology definitely has no effect whatsoever. The research is in many a case qualitative (methods: observations, surveys, and interviews) to evaluate subjective experiences. However, the lack of metrics related to the *health and well-being effects* of the technology is a challenge to meet. There are many traditional tests that can be used to evaluate health condition, and one way to get quantitative research data is to make them electronic (see Chapter 13), but there is plenty of work to be done.

1.9 ETHICAL ASPECTS IN CARE AND TECHNOLOGY

Ethics and social responsibilities related to the current trends in technology and health industry development cover issues such as economy, legislation, and philanthropy. What is legal is not automatically ethical. As to legal aspects alone, laws and regulations vary a lot in different countries. In general, ethical codes and legislation lag behind in the pace of rapid technology development and deployment. Especially issues such as privacy, equality, and safety questions are commonly posed questions. How to assess the truthfulness, reliability, and ethicality of data available? Contrary to our willingness to protect our privacy, we are every day more and more exposed to uncontrolled dissemination of information about ourselves by being involved in social media or using mobile and smart technology (Boyd & Crawford 2011, Carroll & Buchholtz 2015, Howard 2014, Wadhwa 2014).

Certain categories emerge as key ethical issues concerning ethics of technology and care: gains or threats to human good, equity, and affordability of technology. The list of gains or noted benefits in the use of smart technology in a variety of care services from patient's point of view is outstanding: improved independence and self-determination, improved communication between patients and various professional parties involved in care, improved follow-up facilities, improved patient satisfaction, increased social participation, and sense of inclusion. From professionals' viewpoint, the benefits include issues such as cost-effectiveness, increased professional appeal, better service delivery in versatile service environments and contexts, improved monitoring and follow-up of care, service logistics and impacts of care, and enabling individualized but also equity of care supply (Albrecht & Fangerau 2015, Korhonen et al. 2015, Topo 2011, Zwijsen et al. 2011).

Ethical concerns in engaging technology related to care consist most commonly of issues such as technology load, impersonality of care supply and inflexible systems concerning the transfer of responsibility and control between various care suppliers, informational rights, equity and affordability of technology-based services, usability of technology concerned, need of constant user training and support services, and user safety and dignity (Albrecht & Fangerau 2015, Korhonen et al. 2015, Larsen 2012, Zwijsen et al. 2011).

When discussing the ethical topics, another aspect is the ethics and governance related to business and innovations. Recently questions such as ethical, responsible, and sustainable innovations have been discussed more often. UNICEF (2014) has set some generic ethical principles for technology innovations: (1) design with the user, (2) understand the existing ecosystem, (3) design for scale, (4) build for sustainability, (5) be data driven, (6) use open standards, open data, open source, and open innovation, (7) reuse and improve, (8) do no harm, and (9) be collaborative. Also some other frameworks for responsible innovation are presented to assist in reflexion and provide some overall guidelines related to technology innovation and governance (Carroll & Buchholtz 2015, European Commission 2015). Stilgoe et al.'s (2013) four-dimensional model for responsible innovations is presented as a showcase of complexities related to responsible innovations (Table 1.1).

Ethical aspects—like all of the chapters and themes in this book—would require an entire book devoted to each of the topics alone. However, this book tends to provide an overview of the eHealth and eCare context and technology solutions. The book does not try to cover all aspects and introduce all related technology due to the fact that it deals with rapidly developing field and emerging technologies. This book consists of three parts that discuss enabling technology from different points of view. The first section "Opportunities and Barriers of Smart Technology in Care" gives an overview of smart eCare and eHealth technology. The context as well as the needs, possibilities, and challenges of the technology is introduced. This section discusses the technology at a more general level, and detailed information is provided in the next sections. The second section "ICT-Based Platforms and Technology Examples" discusses the technology first at the system level. Both hospital-related platforms and home care systems are introduced and discussed. After introducing platforms, the second part continues with chapters that focus on the device level. Examples on novel sensors are presented to illustrate ongoing developments and further evolving

TABLE 1.1

Four Dimensions of Responsible Innovation

Dimension	Indicative Techniques and Approaches	Factors Affecting Implementation
Anticipation	Foresight	Engaging with existing imaginaries
	Technology assessment	Participation rather than prediction
	Horizon scanning	Plausibility
	Scenarios	Investment in scenario-building
	Vision assessment	Scientific autonomy and reluctance to anticipate
Reflexivity	Multidisciplinary collaboration and training	Rethinking moral division of labor
	Embedded social scientists and ethicists in laboratories	Enlarging or redefining role responsibilities
	Ethical technology assessment	Reflexive capacity among scientists and within institutions
	Codes of conduct	Connections made between research practice and governance
	Moratoriums	
Inclusion	Consensus conferences	Questionable legitimacy of deliberative exercises
	Citizens' juries and panels	Need for clarity about, purposes of and motivation for dialogue
	Focus groups	
	Science shops	Deliberation on framing assumptions
	Deliberative mapping	Ability to consider power imbalances
	Deliberative polling	Ability to interrogate the social and ethical stakes associated with new science and technology
	Lay membership of expert bodies	
	User-centered design	
	Open innovation	Quality of dialogue as a learning exercise
Responsiveness	Constitution of grand challenges and thematic research programs Regulation	Strategic policies and technology *roadmaps*
	Standards	Science-policy culture
	Open access and other mechanisms of transparency	Institutional structure
		Prevailing policy discourses
	Niche management	Institutional cultures
	Value-sensitive design	Institutional leadership
	Moratoriums	Openness and transparency
	Stage-gates	Intellectual property regimes
	Alternative intellectual property regimes	Technological standards

Source: Stilgoe, J., et al., *Research Policy*, 42, 1568–1580, 2013. With permission.

technical solutions. Common commercial sensors are also discussed as part of the platforms. The last section of the book "Case Studies and Field Trials" combines issues discussed in previous sections by introducing case studies and real-life pilots performed. Thus, this book gives an insight into future trends and emerging solutions.

REFERENCES

Albrecht U-V, Fangerau H. Do Ethics Need to Be Adapted to mHelth? A Plea for Developing a Consistent Framework. *World Medical Journal*, July 2015, 61 (2), 72–75.

Aragall F, Neumann P, Sagramola S (with the support of the EuCAN Membership). European Concept of Accessibility—Design for All in Progress, from Theory to Practice. EuCAN—European Concept for Accessibility Network c/o Info-Handicap Luxembourg, 2013.

Boyd D, Crawford K. Six Provocations for Big Data. Paper at Oxford Internet Institute's *A Decade in Internet Time: Symposium on the Dynamics of the Internet and Society* on September 21, 2011. Social Science Electronic Publishing, Inc., 2011. Available at: http://papers.ssrn.com/sol3/papers.cfm?abstract_id=1926431. Retrieved January 21, 2016.

Carroll AB, Buchholtz AK. *Business and Society: Ethics, Sustainability, and Stakeholder Management*. 9th ed. Cengage Learning: Stanford, CA, 2015.

Chan M, Campo E, Estève D, Fourniols J-Y. Smart Homes—Current Features and Future perspectives. *Maturitas*, October 20, 2009, 64 (2), 90–97. Available at: http://www.sciencedirect.com/science/article/pii/S0378512209002606. Retrieved July 06, 2016.

Design for All Foundation. *What is Design for All?*, 2015. Available at: http://designforall.org/design.php. Retrieved July 06, 2016.

Empatica E4 wristband. Available at: https://www.empatica.com/e4-wristband. Retrieved July 06, 2016.

European Commission. *ECHI—European Core Health Indicators*, 2015. Available at: http://ec.europa.eu/health/indicators/echi/list/index_en.htm. Retrieved September 27, 2015.

Flandorfer P. Population Ageing and Socially Assistive Robots for Elderly Persons: The Importance of Sociodemographic Factors for User Acceptance. *International Journal of Population Research*, 2012 (Article ID 829835), 1–13. Available at: http://www.hindawi.com/journals/ijpr/2012/829835/. Retrieved July 06, 2016.

Häyrinen K, Saranto K, Nykänen P. Definition, Structure, Content, Use and Impacts of Electronic Health Records: A Review of the Research Literature. *International Journal of Medical Informatics*, 200877 (5), 291–304. Available at: http://www.sciencedirect.com/science/article/pii/S1386505607001682. Retrieved July 06, 2016.

Heikenfeld J. Sweat Sensors Will Change How Wearables Track Your Health. *IEEE Spectrum*, 2014. Available at: http://spectrum.ieee.org/biomedical/diagnostics/sweat-sensors-will-change-how-wearables-track-your-health. Retrieved December 30, 2015.

Howard A. Disruptive Technologies Pose Difficult Ethical Questions for Society. *TEchRepublic*, April 22, 2014. Available at: http://www.techrepublic.com/article/disruptive-technologies-pose-difficult-ethical-questions-for-society/. Retrieved January 20, 2016.

Huuhtanen K. *Puhetta tukevat ja korvaavat kommunikointimenetelmät Suomessa*. Kehitysvammaliitto ry.: Helsinki, Finland, 2012.

Karlen W. *Mobile Point-of-Care Monitors and Diagnostic Device Design*. CRC Press, Boca Raton, FL, 2014.

Kemppainen E. Kohti esteetöntä yhteiskuntaa—Yhteiskuntapolitiikan normatiiviset keinot esteettömyyden edistämisessä. Stakesin raportteja 33/2008. Helsinki, Finland, 2008.

Kiili K, Merilampi S. Developing Engaging Exergames with Simple Motion Detection. *Proceedings of the 14th International Academic MindTrek Conference: Envisioning Future Media Environments*, ACM, NY, pp. 103–110, 2010.

Knaup P, Schöpe L. Using Data from Ambient Assisted Living and Smart Homes in Electronic Health Records. *Methods of Information in Medicine*, 2014, 53(3), 149–151. Available at: http://www.ncbi.nlm.nih.gov/pubmed/24828122. Retrieved July 06, 2016.

Koivisto A, Merilampi S, Kiili K. Mobile Exergames for Preventing Diseases Related to Childhood Obesity. 4th International Symposium on Applied Sciences in Biomedical and Communication Technologies (ISABEL), Barcelona Spain, 2011.

Koivisto A, Merilampi S, Kiili K, Sirkka A, Salli J. Mobile Activation Games for Rehabilitation and Recreational Activities—Exergames for the Intellectually Disabled and the Older Adults. *Journal of Public Health Frontier*, 2013, 2 (3), 122–132.

Koivisto A, Merilampi S, Sirkka A. Mobile Games Individualise and Motivate Rehabilitation in Different User Groups. *International Journal of Game-Based Learning*, April 2015, 5 (2), 1–17.

Koivisto A, Merilampi S, Sirkka A. Mobile Rehabilitation Games—User Experience Study. European Conference on Games Based Learning, ECGBL, Berlin, October 2014.

Korhonen E-S, Nordman T, Eriksson K. Technology and Its Ethics in Nursing and Caring Journals: An Intregrative Literature Review. *Nursing Ethics*, 2015, 22 (5), 561–576.

Kuusi O. *Ikääntyneiden itsenäistä selviytymistä tukeva tulevaisuuspolitiikka ja geronteknologia. Geronteknologia-arvioinnin loppuraportti*. Eduskunnan kanslian julkaisu 7/2001. Edita: Helsinki, Finland.

Larsen A-C. Trappings of Technology: Casting Palliative Care Nursing as Legal Relations. *Nursing Inquiry*, 2012, 19 (4), 334–344.

Lindeman D. *Technologies to Help Older Adults Maintain Independence: Advancing Technology Adoption*. Center for Technology and Aging Briefing Paper, Oakland, CA, 2009. Available at http://www.techandaging.org/briefingpaper.pdf. Retrieved July 06, 2016.

Löfqvist C, Nygren C, Széman Z, Iwarsson S. Assistive Devices Among Very Old People in Five European Countries. *Scandinavian Journal of Occupational Therapy*, 2005, 12, 181–192.

Mellors WJ. Design for All. In *eAccessibility*. Lehne PH (ed.). Vol. 100, No. 1. Telenor ASA, Norway, 2004. Available at: https://www.telenor.com/wp-content/uploads/2012/05/T04_1.pdf. Retrieved July 06, 2016.

Merilampi S, Sirkka A, Leino M, Koivisto A, Finn E. Cognitive Mobile Games for Memory Impaired Older Adults. *Journal of Assistive Technologies*, November 2014, 8 (4), 207–223.

Phillips J. The 10 most likely sensors in a 10-sensor Apple smartwatch, PC World, 2014. Available at: http://www.pcworld.com/article/2366126/the-10-most-likely-sensors-in-a-10-sensor-apple-smartwatch.html. Retrieved December 30, 2015.

Pieper M, Antona M, Cortés U. Ambient Assisted Living. *ERCIM News*, October 2011, 87. Available at: http://www.ics.forth.gr/files/publications/antona/2011/Pieper_et_al.pdf. Retrieved July 06, 2016.

Salminen A-L. *Apuvälinekirja*. Kehitysvammaliitto ry.: Helsinki, Finland, 2010.

Sirkka A, Merilampi S, Koivisto A, Leinonen M, Leino M. User Experiences of Mobile Controlled Games for Activation, Rehabilitation and Recreation of the Elderly and Physically Impaired. pHealth conference, Porto, Portugal, 2012.

Stilgoe J, Owen R, Macnaghten P. Developing a Framework for Responsible Innovation. *Research Policy*, 2013, 42, 1568–1580.

Suhonen L, Siikanen T. *Hyvinvointiteknologia sosiaali- ja terveysalalla: hyöty vai haitta?*. Lahden ammattikorkeakoulu: Lahti, Finland, 2007.

Topo P. Ethical Issues and Welfare Technology. Ministry of Social Affairs and Health in Finland, ETENE, 2011. Available at: http://www.nordicwelfare.org/PageFiles/599/P%C3%A4ivi%20Topo%20-%20Ethical%20issues%20and%20welfare%20technology.pdf. Retrieved January 28, 2016.

UNICEF 2014. Principles for Innovation and Technology Development. Available at: http://www.unicef.org/innovation/innovation_73239.html. Retrieved January 22, 2016.

Väyrynen S, Kirvesoja H. Johdatus geronteknologaan. *Geronteknologian perusteita ja sovellutuksia. Työtieteen jaoksen hankeraportteja.* Teoksessa Oikarinen A, Sinisammal J, Tornberg V, ja Väyrynen S (toim.). Vol. 4. s. 5–10. Oulun Yliopisto, Oulu, Finland, 1998.

Wadhwa V. Laws and ethics can't keep pace with technology: Codes we live by, laws we follow, and computers that move too fast to care, *MIT Technology Review*, 2014. Available at: http://www.technologyreview.com/view/526401/laws-and-ethics-cant-keep-pace-with-technology/. Retrieved January 20, 2016.

Wichert R, Klausing H. Ambient Assisted Living, 6. AAL-Kongress 2013 Berlin, Germany, January 22–23, 2013, 2008. Available at: http://link.springer.com/book/10.1007/978-3-642-37988-8. Retrieved July 06, 2016.

Zwijsen S, Niemeijer AR, Hertogh CMPM. Ethics of Using Assistive Technology in the Care for Community-dwelling Elderly People: An Overview of the Literature. *Aging & Mental Health*, May 2011, 15 (4), 419–427.

http://www.papunet.net. Retrieved December 14, 2015.

http://www.thl.fi/apuvälineet. Retrieved December 14, 2015.

http://www.tikoteekki.fi. Retrieved December 14, 2015.

2 Drivers and Trends in Technology Deployment in Care Services

Andrew Sirkka

CONTENTS

2.1 INTRODUCTION

Digitalization expedites a shift from intramural care toward seamless and even ubiquitous provision of services. The drivers for this transition consists of efforts to more cost-efficient care delivery, intelligent analysis of patient information, more scalable software delivery models, and new types of interaction between individuals and caregivers using modern technologies and online service platforms.

Some of the most commonly recognized transition drivers are as follows: globalization, technology development, digitalization, interdisciplinary research and development, changes in life styles and human right policies, education, economy, and transitions in service structures.

Although the European Union (EU) has drawn common strategies on healthcare transitions, each EU member state still has its own national service system. Only to mention a couple of examples, the Innovation Union strategy aims to maximize the EU's capacity for innovation and research and channel it toward societal challenges by promoting active and healthy aging, and the Digital Agenda for Europe strategy focuses on developing and using digital applications aiming at improving the quality of care, reducing medical costs, and fostering independent living among people (European Commission 2015b).

This chapter provides an overview of the current drivers and trends in quite an intense transition that has taken place recently in the healthcare industry and services in Europe. More detailed information on a variety of advanced technologies applied in care delivery is presented in Chapters 3–9.

2.2 DEMOGRAPHICS AND ECONOMICS

As to demographic drivers, the most commonly discussed concern is the aging structure of the European population and the evolution of its health status. The number of aging population is automatically expected to increase the public healthcare expenditure. However, a relatively large number of studies indicate that the ageing population is taking more active role in their own health, and, thanks to their own efforts, the ability to remain longer active, self-supportive, and in good health is rapidly improving. Active and healthy aging policies are based on the evidence that physically, socially, and mentally active older adults are also healthier older adults. Active aging studies indicate the expenditure curve progressively postponing the age-related increases in expenditure, which is why calculating pure demographic statistics with implicit assumptions of age-related health status could be misleading (Chang et al. 2013; Cyarto et al. 2012; Flöel 2010; Hogan 2013; Kamegaya 2012). What actually matters therefore is not the very aging, but the prevalence of severe illnesses in the latter part of human life span and the expenditure related to the care close to death (so called death-related costs) (de la Maisonneuve & Oliveira Martins 2013; Pammolli et al. 2008; Saraceno 2010).

Technological development and deployment of more advanced technology in medicine and healthcare in general has increased the expenses, but also improved the quality of care in terms of improved medical and operational efficacy, and patient safety and satisfaction. Cost-effectiveness is identified as the most important investment criterion. Healthcare systems globally have similarities in their governance and decision-making processes, such as political governance, heavy and very structured management, information, and operational systems, that tend to increase the expenditure (Reiter & Song 2011; Sorenson et al. 2013; Wernz et al. 2014).

Health inequities, even if being a pressure point in strategies for years, are stubbornly pointing out even in the latest statistics. One of the explanations is the global economic downturn that is seen to have a profound importance for the health and well-being of populations. On the other hand, the economic downturn in Europe is not visible as convergence of service provision, but it appears mainly as increased reliance on private healthcare financing. European countries have achieved significant gains in population health, but large inequalities in health status both across and within countries remain. The gap between countries with the highest and lowest life expectancies remains around 8 years. In addition, there are also persistently large inequalities within countries among people from different socio-economic groups, with individuals with higher levels of education and income enjoying better health and living several years longer than those more disadvantaged. These disparities are linked to many factors, including some outside healthcare systems, such as the environment in which people live, individual lifestyles and behaviors, and differences in access to and quality of care (Montanari & Nelson 2013; OECD 2014).

The European Core Health Indicator (ECHI) is a set of 88 indicators developed by European Commission to create a comparable health information and knowledge system to monitor health at the EU level (European Commission 2015a). Based on the latest data published by OECD in 2014, inequities, accessibility to care, and safety measures are still as the centerpiece of development among the EU member states.

Social determinants of health contribute to producing other social benefits such as well-being, improved education, more sustainable communities, and improved social cohesion and integration. Early-years skills are crucial to self-esteem, and long-term health and well-being. Therefore, investment on the social determinants of health demonstrates can directly contribute to attaining other sectoral and government goals and challenges the notion that health drains public resources (UCL 2011).

2.3 HUMAN RIGHTS AND LIFESTYLE CHANGES

Human rights are widely recognized as the universal and indivisible part of human life. The EU has declared a strong engagement to promote and protect human rights, democracy, and rule of law worldwide. The EU Strategic Framework and Action Plan on Human Rights and Democracy since June 2012 has continued to be the reference document setting out the guiding principles and main priorities in EU policies. Every year the European Council adopts its Annual Report on Human Rights and Democracy to assess the actions taken to address the Action Plan's priorities and mapping in detail human rights situation worldwide. Citizens' own responsibilities in regard to maintaining health is emphasized in the EU health strategies. Call for life style changes in many aspects has been announced recently in order to stay within the limits of sustainability starting from a redefinition of progress, taking into account environmental and social indicators as well as economic growth (Gray et al. 2012; McCormick 2011; Rider & Makela 2003; Schirmer & Michailakis 2012).

Community participation as citizens' active involvement in formal or informal activities, programs and/or discussions to bring about a planned change or improvements in community life, services, and/or resources has evidenced a positive impact in local health and sustainable development. More recently, the idea of a *whole-of-society* approach to achieving health goals has emerged. It is based on the idea that civil society has a vital contribution to make to an interconnected health system (Figure 2.1; Public Health England 2015).

Technology has been notified as one of the key means to improve individual's participation and possibilities in the society. Recent studies clearly show that population's ability to use technology has improved, and the use of information and communication technology (ICT) increased rather rapidly in all age groups. The Internet has become an important means for daily life, education, work, and participation in society, enabling people to access information and services at any time from any place. Most Internet users search for information and news, consult wikis, participate in social networks, and buy products online. In 2013, two in five individuals

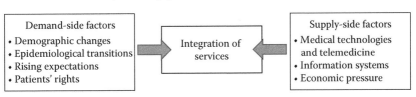

FIGURE 2.1 Driving forces for integration of services. (Adapted from Gröne, O. and Garcia-Barbero, M., *International Journal of Integrated Care*, 1, 2001.)

aged 16–74 years in the EU contacted or interacted with public authorities or public services via websites for private purposes. Income tax declaration was a major reason for use among those who used e-government services (Care Innovations 2013; Damant & Knapp 2015; Seybert & Reinecke 2013). The Internet habits of age groups, socio-economic groups, and different countries vary significantly from zero use to 61% of Internet users using it for e-shopping. Only a fifth of the EU population have never used the Internet (Seybert & Reinecke 2013).

A person's attitude has an outstanding impact on the ICT use. Negative attitudes and anxieties mostly derive from diffidence of using the equipment correctly, concerns of getting the software blocked by using errors or computer viruses, fears related to abuse of personal information on the Internet, and other bad experiences or uncertainties. In general, men experience less computer-related anxiety and have more positive attitudes than women, although the gender gap in the ICT use appears to be rapidly narrowing. In Britain, 55% of female and 64% of male respondents used the Internet in 2003, whereas in 2013, the percentages were increased up to 78% and 79%. As to the Internet self-efficacy, 70% of women and 77% of men rated their Internet ability as excellent or good (Damant & Knapp 2015; Dutton & Blank 2013).

Social inclusion is one of the key concepts in the European social policy. According to European Commission (2013), nearly one quarter of the EU's population is considered to be at risk of poverty or social exclusion. Social inclusion is not only about successful labor-market participation. It is also about the maintenance of wellness throughout life. Aging of population calls for new ways of empowering individuals to stay active in working life and social participation in its all aspects. Interestingly, digital games and gamification in general seem to provide a novel means to enhance integration of disengaged and disadvantaged learners into society, to raise awareness about health issues, to promote health and well-being and raise awareness about important social policy issues, such as equity and poverty to enable participatory community planning (Silver 2010).

2.4 TRENDS TRANSFORMING HEALTHCARE SUPPLY SYSTEMS

A variety of trends and drivers having an impact on and transforming healthcare have been identified recently. According to Shah (2015), there are three trends: cloud-based electronic health records, patient-centric devices, and Big Data analytics and patient access. White (2014) points out four trends: (1) patient convenience in data handling, (2) rise of patient apps, (3) lowering costs through telehealth, and (4) innovative care models for large-scale consumers of health services.

Accenture (2015), on the other hand, names five information technology trends: (1) "The Internet of Me"—healthcare experience becomes more personalized and more convenient for patients, (2) the Internet of Things—healthcare disruptive technology produces highly connected hardware that increases access to care, patient engagement, and clinical collaboration, (3) "The Platform (R)evolution"—smarter healthcare platforms provide patients with real-time health information and more self-monitoring through mobile devices, (4) Cognitive Tech, Big Data, and Health Analytics—technology could help diagnose a rash, confirm a fever, and grant patients greater access to their medical records, and (5) Workforce Transformation—technology enables providers to be more efficient, handle more complex tasks, and improve patient care.

Holoubek (2015) identifies ten trends transforming healthcare: (1) organizational maturity, (2) data science movement, (3) data liquidity, (4) data security, (5) humans first thinking, (6) new models, (7) new care team, (8) new investment insights, (9) open innovation, and (10) clinical innovation.

Based on the historical development of technologies used in care provision, which has been technology or producer-driven for most of the time, more attention has to be paid to the needs and wishes of end-users (customers, care professionals, and (in)formal caregivers) to make services and incorporated devices purposeful, functional, accessible, and easy to use. On one hand, smart technology enables more individualized and more independent ways of working, on the other hand, it requires new skills and competences in care delivery and urges for new tools and measurements to be used for quality assurance, including patient reported outcomes. To add one apt example, think of the constant flow of customer information coming in that requires reorganization of professional care supply across the whole set of competences (attitude, knowledge, and skills).

Technology development toward assistive and enabling technologies, user-friendly, and automated positive technology has been outstanding recently. Positive technology (Figure 2.2) indicates to technology used for improving the quality of our personal experience, suggesting specific strategies to modify/improve each of the different dimensions involved while generating motivation and engagement in the process. The use of positive technology tools and strategies allows the expansion of healthcare beyond the traditional intramural healthcare services. Today's healthcare is increasingly using intelligent tools to obtain smarter clinical information to improve patient outcomes. Deployment of advanced simulation technologies such as virtual or augmented reality and spontaneous peer networks would strongly change the healthcare service supply. With more information becoming available, the paradigm shift toward preventive measure could remold the healthcare from episodic, isolated patient care to electronic patient-centered care (Graffigna et al. 2013; McCullagh 2012; Nasi et al. 2015; Sorenson et al. 2013; Wu et al. 2011; Ziefle et al. 2010).

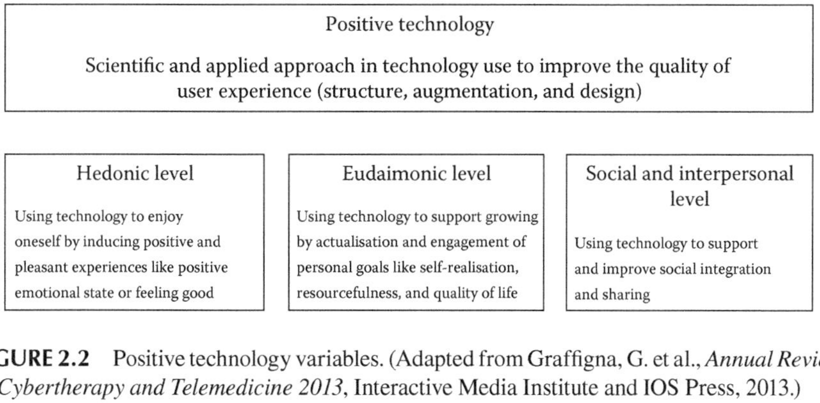

Positive technology
Scientific and applied approach in technology use to improve the quality of user experience (structure, augmentation, and design)

Hedonic level	Eudaimonic level	Social and interpersonal level
Using technology to enjoy oneself by inducing positive and pleasant experiences like positive emotional state or feeling good	Using technology to support growing by actualisation and engagement of personal goals like self-realisation, resourcefulness, and quality of life	Using technology to support and improve social integration and sharing

FIGURE 2.2   Positive technology variables. (Adapted from Graffigna, G. et al., *Annual Review of Cybertherapy and Telemedicine 2013*, Interactive Media Institute and IOS Press, 2013.)

2.5 SERVICE DESIGN AND INNOVATIVE SERVICES

As discussed above, the delivery of care services is changing due to demographic and societal trends. In Europe, healthcare policy promotes inclusion and enablement of citizens to take ownership and responsibility in their own health issues, and, for example, the UN convention on the Rights of Persons with Disabilities also demands deinstitutionalization of care services. The pressure toward more integrated and flexible service models is increasing.

Service delivery redesign requires new grounds for sustainable, cost-effective service infrastructure, embedded multiprofessional approaches, engagement of smart technology, assurance of safe and secure services in multiple, and also extramural environments in line with ethical and legal aspects (Jaakkola & Alexander 2014). Innovative services require intensified client/patient engagement in order to achieve the positive and empowering atmosphere in care supply (Figure 2.3).

Consumer- or client-based health information technology is increasingly taken into use in various healthcare services. The effectiveness of these technologies indicates the importance of multilateral communication in assessment of current patient status, interpretation of information regarding setting the goals in tailored care plans, and means available. Deployment of ICT systems challenges healthcare service systems to shift the focus of care away from clinical appointments toward client empowerment through multidisciplinary team activities. Therefore, healthcare professionals are also challenged to attain new competences. That is why the concept technology competency has turned to play an increasing role in all healthcare professionals' competencies (Barakat et al. 2013; Chia 2002; Dünnebeil 2012; Jimison et al. 2008; Kocher & Sahni 2011; Tax et al. 2012).

Another critical factor is dragging healthcare into the Digital Age. Technology in care has already created some new service concepts such as personalized health (p-health), mobile health (m-health), or telehealth (also called eHealth). Remote care (eHealth or telehealth), online services in healthcare, and social sector and other forms of innovative and multiprofessional services are means to provide care in accessible and user-driven ways. Telehealth, the provision of care at a distance, is a key component in future integrated care. For some reasons, eHealth or telecare is more progressed in higher-income countries even if WHO (2015) has supported development of eHealth or telehealth in low- and middle-income countries to improve the accessibility and quality of care (Mair et al. 2012; Wootton et al. 2012).

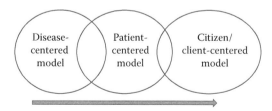

FIGURE 2.3 Paradigm shift toward innovative care. (Adapted from Graffigna, G. et al., *Annual Review of Cybertherapy and Telemedicine 2013*, Interactive Media Institute and IOS Press, 2013.)

Information systems as means for communication between different stakeholders (for example, healthcare professionals, customers, local authorities, and health assurance companies) should more effectively be deployed. Efficient service supply demands excessive ability in multiprofessional communication in interactive and shared expertise (common understanding of and engagement in different specific professional expertise to reach the common goal). All this force the service providers to face the challenges due to digitalization of everything. Integrated care services require multidisciplinary teamwork in order to deliver a coordinated response to each individual's care requirements. The biggest challenges are the loss of control over the customer relationship, increased competition and commoditization, and the need to continuously engage with customers, suppliers, partners, and employees. The generation Y (digital natives born after 1980) is not only accelerating the take-up of new digital technology but also posing additional challenges with their expectations and demands representing over a quarter of the world's population (Ernst & Young 2011; Mair et al. 2012; Stroetmann et al. 2010).

mHealth is stated to revolutionize the healthcare service delivery leading to better outcomes, decreased costs, and improved sustainability of the healthcare system. At the end of 2013, there were 3 million patients using connected home medical monitoring devices worldwide. There is a strong trend toward incorporating more connectivity in medical devices in order to enable new services and value propositions. Furthermore, connectivity is gaining momentum in several other device categories such as glucose monitoring and air flow monitoring, as well as enabling the emergence of new categories such as telehealth hubs and medication compliance monitoring solutions.

The adoption of remote patient monitoring solutions is driven by a wide range of incentives, related to everything from demographics and technology development to new advancements in medical treatment. However, there are a number of barriers, including resistance to change among healthcare organizations and clinicians, misaligned incentive structures and the financing of wireless solutions by what is at large an underfunded healthcare sector. Several catalysts are nevertheless speeding up the rate of adoption—in particular incentives from payers and insurance companies, national health systems that support remote monitoring, and a shift to performance-based payment models.

ICT accessibility for persons with disabilities should be a priority for all service providers. ICT accessibility means removing barriers so that all individuals, including persons with visual, hearing, mobility, dexterity, and cognitive disabilities, can use it. Despite advances that have taken place in recent years, persons with disabilities continue to face barriers in using ICTs. Although already some service providers support online services in various ways, there is still a lot to do in this field to make it accessible for all.

In increasing pace, some innovative health services have emerged, utilizing new technologies but also service designing principles. Gaming has opened novel approaches in patient education, health promotion, skills development, and rehabilitation. Modern communication technologies are deployed in healthcare to provide call-in services giving people an affordable 24-hour access to doctor consultations and other health services. Also some private, walk-in kiosks for patient care have been established. Because of economic downturn also the issues related to the access and affordability to top-quality health products and services should be paid attention to. Giving individuals incentives by insurance companies to maintain health has gained good results, allowing members to reward individuals for taking responsibility for their own health. Applying innovative and integrated services provides completely novel and even more affordable healthcare concept that requires reconsideration and reconstruction of prevailing service systems (Fast Company 2013; Kocher & Sahni 2011; Macaulay et al. 2012; Thadani 2015; van der Eijk et al. 2013).

In most cases, change is good. Successful changes require good planning, reassessment of what is the core of intended services, and to make preparations in terms of researching options, adapting to changes in client behaviors, and educating people. In general, what is very much needed at present is the courage to shake off the conventional thinking blocking the constructive development both in service models and professional qualifications.

REFERENCES

Accenture. 2015. Healthcare IT: Top 5 eHealth trends reshaping the industry in 2015. Available at: https://www.accenture.com/us-en/insight-healthcare-technology-vision-2015.aspx. Retrieved September 17, 2015.

Barakat A, Woolrych RD, Sixsmith A, Kearns WD, & Kort HSM. 2013. eHealth technology competencies for health professionals working in home care to support older adults to age in place: Outcomes of a two-day collaborative workshop. *Medicine 2.0*, 2(2), e10. Available at: http://www.medicine20.com/2013/2/e10. Retrieved October 12, 2015.

Care Innovations. 2013. Older populations have adopted technology for health. *White Paper*. Available at: http://www.careinnovations.com/wp-content/uploads/filebase/Older-Populations-Adopt-Health-Technology.pdf. Retrieved September 25, 2015.

Chang Y-K, Huang C-J, Chen K-F, & Hung T-M. 2013. Physical activity and working memory in healthy older adults: An ERP study. *Psychophysiology*, 50(11), 1174–1182.

Chia CY. 2002. Identifying information technology competencies needed in Singapore nursing education. *Computeres, Informatics, Nursing*, 20(5), 209–214.

Cyarto EV, Lautenschlager NT, Desmond PM, Ames D, Szoeke C, Salvado O, Sharman MJ, Ellis KA, Phal PM, Masters CL, Rowe CC, Martins RN, & Cox KL. 2012. Protocol for a randomized controlled trial evaluating the effect of physical activity on delaying the progression of white matter changes on MRI in older adults with memory complaints and mild cognitive impairment: The AIBL Active trial. *BMC Psychiatry*, 12(1), 167–177.

Damant J & Knapp M. 2015. What are the likely changes in society and technology which will impact upon the ability of older adults to maintain social (extra-familial) networks of support now, in 2025 and in 2040? Future of an aging population: Evidence review. Personal Social Services Research Unit London School of Economics and Political Science. Foresight, Government Office for Science. Available at: https://www.gov.uk/government/uploads/system/uploads/attachment_data/file/440191/gs-15-6-technology-and-support-networks.pdf. Retrieved September 25, 2015.

de la Maisonneuve C & Oliveira Martins J. 2013. Public spending on health and long-term care: A new set of projections. OECD Economic Policy Papers No 6. OECD Publishing, Paris.

Dünnebeil S, Sunyaev A, Blohm I, Leimeister JM, & Krcmar H. 2012. Determinants of physicians' technology acceptance for e-health in ambulatory care. *International Journal of Medical Informatics*, 81(11), 746–760.

Dutton, WH & Blank G. 2013. *Cultures of the Internet: The Internet in Britain*. Oxford Internet Survey 2013 Report. Available at: http://oxis.oii.ox.ac.uk/wp-content/uploads/2014/11/OxIS-2013.pdf. Retrieved September 7, 2015.

Ernst & Young LLP. 2011. *The Digitalisation of Everything: How Organisations Must Adapt to Changing Consumer Behaviour.* Ernst & Young: London.

European Commission. 2013. *The Potential of Digital Games for Empowerment and Social Inclusion of Groups at Risk of Social and Economic Exclusion: Evidence and Opportunity for Policy.* JRC Scientific and Policy Reports, EUR25900 EN. Available at: https://ec.europa.eu/jrc/en/publication/eur-scientific-and-technical-research-reports/potential-digital-games-empowerment-and-social-inclusion-groups-risk-social-and-economic. Retrieved September 25, 2015.

European Commission. 2015a. *ECHI—European Core Health Indicators.* Available at: http://ec.europa.eu/health/indicators/echi/list/index_en.htm. Retrieved September 27, 2015.

European Commission. 2015b. *Europe 2020—For a Healthier EU.* Available at: http://ec.europa.eu/health/europe_2020_en.htm. Retrieved September 17, 2015.

Fast Company. 2013. *The World's Top 10 Most Innovative Companies in Health Care.* Available at: http://www.fastcompany.com/most-innovative-companies/2013/industry/healthcare. Retrieved October 7, 2015.

Flöel A, Ruscheweyh R, Krüger K, Willemer C, Winter B, Völker K, Lohmann H, Zitzmann M, Mooren F, Breitenstein C, & Knecht S. 2010. Physical activity and memory functions: Are neurotrophins and cerebral gray matter volume the missing link? *Neuroimage*, 49(3), 2756–2763.

Graffigna G, Barello S, Wiederhold BK, Bosio AC, & Riva G. 2013. Positive technology as a driver for health engagement. In: B.K. Wiederhold and G. Riva (Eds). *Annual Review of Cybertherapy and Telemedicine 2013. Studies in Health Technology and Informatics.* Vol. 191, pp. 9–17. Interactive Media Institute and IOS Press: Amsterdam, the Netherlands. Available at: http://www.academia.edu/3892668/Positive_Technology_as_a_Driver_for_Health_Engagement. Retrieved September 2, 2015.

Gray BM, Weng W, & Holmboe ES. 2012. An assessment of patient-based and practice infrastructure-based measures of the patient-centered medical home: Do we need to ask the patient? *Health Services Research*, 47(1), part 1, 4–21.

Gröne O & Garcia-Barbero M. 2001. Integrated care—A position paper of the WHO European office for integrated health care services. *International Journal of Integrated Care*, 1, 1–10. Available at: http://www.ijic.org/articles/10.5334/ijic.28/. Retrieved September 2, 2015.

Hogan CL. 2013. Exercise holds immediate benefits for affect and cognition in younger and older adults. *Psychology and Aging*, 28(2), 587–594.

Holoubek S. April 2015. 10 trends transforming healthcare in 2015. *Medical, Marketing & Media Magazine.* Available at: http://www.mmm-online.com/issue/april/02/2015/2431/ Retrieved September 17, 2015.

Jaakkola E & Alexander M. 2014. The role of customer engagement behavior in co-creation: A service system perspective. *Journal of Service Research*, 17(3), 1–15.

Jimison H, Gorman P, Woods S, Nygren P, Walker M, Norris S, & Hersh W. 2008. *Barriers and Drivers of Health Information Technology Use for the Elderly, Chronically Ill, and Underserved.* Evidence Report/Technology Assessment No. 175 (Prepared by the Oregon Evidence-based Practice Center under Contract No. 290-02-0024). AHRQ Publication No. 09-E004. Rockville, MD: Agency for Healthcare Research and Quality. November 2008.

Kamegaya T, Maki Y, Yamagami T, Yamaguchi T, Murai T, & Yamaguchi H. 2012. Pleasant physical exercise program for prevention of cognitive decline in community-dwelling elderly with subjective memory complaints. *Geriatrics & Gerontology International*, 12(4), 673–679.

Kocher R & Sahni RN. 2011. Rethinking health care labor. *The New England Journal of Medicine*, 365(15), 1370–1372.

Macaulay LA, Miles I, Wilby J, Tan YL, Zhao L, & Theodoulidis B (Eds). 2012. *Case Studies in Service Innovation*. Springer: New York.

Mair FS, May C, O'Donnell C, Finch T, Sullivand F, & Murraye E. 2012. Factors that promote or inhibit the implementation of e-health systems: An explanatory systematic review. *Bulletin of the World Health Organization*, 90, 357–364. Available at: http://www.who.int/bulletin/volumes/90/5/en/. Retrieved September 24, 2015.

McCormick A. 2011. Self-determination, the right to die, and culture: A literature review. *Social Work*, 56(2), 119–128.

McCullagh PJ, Mountain GA, Blacka ND, Nugenta CD, Zhenga H, Daviesa RJ, Galwaya L, Hawleyb MS, Mawsonc SJ, Wrighte P, Ecclestond C, Nasrc N, & Parker SJ. 2012. Knowledge transfer for technology based interventions: Collaboration, development and evaluation. *Technology and Disability*, 24, 233–243.

Montanari I & Nelson K. 2013. Social service decline and convergence: How does healthcare fare? *Journal of European Social Policy*, 23(1), 102–116.

Nasi G, Cucciniello M, & Guerrazzi C. 2015. The role of mobile technologies in health care processes: The case of cancer supportive care. *Journal of Medical Internet Research*, 17(2), e26. Available at: http://www.ncbi.nlm.nih.gov/pmc/articles/PMC4342745/. Retrieved February 24, 2015.

OECD. 2014. *Health at a Glance: Europe 2014*. OECD Publishing: Paris, France. Available at: http://dx.doi.org/10.1787/health_glance_eur-2014-en. Retrieved September 27, 2015.

Pammolli F, Riccaboni M, & Magazzini L. 2008. *The Sustainability of European Health Care Systems: Beyond Income and Ageing*. IMT Institute For Advanced Studies: Lucca, Italy. Available at: http://mpra.ub.uni-muenchen.de/16026/ MPRA Paper No. 16026.

Public Health England. 2015. *A Guide to Community-centred Approaches for Health and Wellbeing*. PHE Publications: London. Available at: https://www.gov.uk/government/uploads/system/uploads/attachment_data/file/402889/A_guide_to_community-centred_approaches_for_health_and_wellbeing__briefi___.pdf. Retrieved September 25, 2015.

Reiter KL & Song PH. 2011. The role of financial market performance in hospital capital investment. *Journal of Healthcare Finance*, 37(3), 38.

Rider ME & Makela CJ. 2003. A comparative analysis of patients' rights: An international perspective. *International Journal of Consumer Studies*, 27(4), 302–315.

Saraceno C. 2010. Social inequalities in facing old-age dependency: A bi-generational perspective. *Journal of European Social Policy*, 20(1), 32–44.

Schirmer W & Michailakis D. 2012. The latent function of "responsibility for one's health" in Swedish healthcare priority-setting. *Health Sociology Review*, 21(1), 36–46.

Seybert H & Reinecke P. 2013. Internet use statistics—individuals. *Statistics in Focus 29/2013*. Eurostat. Available at: http://ec.europa.eu/eurostat/statistics-explained/index.php/Internet_use_statistics_-_individuals. Retrieved September 25, 2015.

Shah M. 2015. 3 Technology trends transforming health care. *Forbes*, April 13, 2015. Available at: http://www.forbes.com/sites/athenahealth/2015/04/13/3-technology-trends-transforming-health-care/. Retrieved August 29, 2015.

Silver H. 2010. Understanding social inclusion and its meaning for Australia. *Australian Journal of Social Issues*, 45(2), 183–211.

Sorenson C, Drummond M, & Bhuiyan Khan B. 2013. Medical technology as a key driver of rising health expenditure: Disentangling the relationship. *ClinicoEconomics and Outcomes Research*, 5, 223–234.

Stroetmann KA, Kubitschke L, Robinson S, Stroetmann V, Cullen K, & McDaid D. 2010. How can telehealth help in the provision of integrated care? World Health Organization on behalf of the European Observatory on Health Systems and Policies, Policy brief 13.

Tax SS, McCutcheon D, & Wilkinson IF. 2012. The service delivery network (SDN). A customer-centric perspective of the customer journey. *Journal of Service Research*, 16(4), 454–470.

Thadani T. 2015. The digital doctor is in: Next wave in health care. *USA Today*, July 8, 2015.

UCL. 2011. *The Future of Healthcare in Europe*. UCL European Institute: London.

van der Eijk M, Faber MJ, Aarts JWA, Kremer JAM, Munneke M, & Bloem BR. 2013. Using online health communities to deliver patient-centered care to people with chronic conditions. *Journal of Medical Internet Research*, 15(6), e115.

Wernz C, Zhang H, & Phusavat K. 2014. International study of technology investment decisions at hospitals. *Industrial Management & Data Systems*, 114(4), 568–582. Available at: http://www.msdm.ise.vt.edu/Paper/Wernz_et_al_2014_International_Study_of_Technology_Investment_Decisions_at_Hospitals_IMDS.pdf. Retrieved September 2, 2015.

White J. 2014. 4 Healthcare trends to watch in 2015. *Healthcare Business & Technology*, December 23, 2014. Available at: http://healthcarebusinesstech.com/healthcare-trends-2015/. Retrieved September 7, 2015.

WHO. 2015. *Health Topics*. Available at: http://www.who.int/topics/ehealth/en/. Retrieved September 24, 2015.

Wootton R, Geissbuhler A, Jethwani K, Kovarik C, Person DA, Vladzymyrskyy A, Zanaboni P, & Zolfo M. 2012. Long-running telemedicine networks delivering humanitarian services: Experience, performance and scientific output. *Bulletin of the World Health Organization*, 90, 341–347. Available at: http://www.who.int/bulletin/volumes/90/5/en/. Retrieved September 2, 2015.

Wu G, Talwar S, Johnsson K, Himayat N, & Johnson KD. 2011. M2M: From mobile to embedded Internet. *IEEE Communications Magazine*, 49(4), 36–43.

Ziefle M, Röcker C, Kasugai K, Klack L, Jakobs E-M, Schmitz-Rode T, Russell P, & Borchers J. 2010. eHealth—Enhancing mobility with aging. In: M. Tscheligi, B. de Ruyter, J. Soldatos, A. Meschtscherjakov, C. Buiza, W. Reitberger, N. Streitz, T. Mirlacher (Eds). *Roots for the Future of Ambient Intelligence, Adjunct Proceedings of the Third European Conference on Ambient Intelligence (AmI'09)*. November 18–21, 2009, Salzburg, Austria, pp. 25–28.

3 Approaches to Smart Technology Deployment in Care

Enda Finn and John Loane

CONTENTS

3.1 INTRODUCTION

This chapter explores how we might best approach the deployment of *usable* and also *appropriate* smart technologies, in particular smaller scale-connected software applications or apps typically combined with various devices or sensors, in a care setting. *Well-being enhancement* should be the ultimate goal of deploying any smart

31

technology in a care setting. Ideally, such enhancements need to be specific, tangible, and also measurable, addressing either a particular well-being attribute or an interdependent group of attributes, for example, physical dexterity, cognitive functioning, or perhaps even mood improvement for either a single individual or group of individuals. At its simplest, the term *deployment* can be understood as providing an *effective* solution to a particular real-world problem. The smart technology deployed can be said to be effective where first it works, meaning it functions correctly and reliably; second, it is deployed within the normal and necessary constraints of cost and time; and third, it can be seen to provide some measures of improvement, enhancement, or support.

It may be that an existing smart technology is simply identified, adapted, and then delivered as a potential solution in a specific case. More likely though is where at least some degree of investigation, requirements gathering, design, development, and testing of either new hardware and/or software is needed or intended. Therefore, deployment will usually involve all the stages within a technology development life cycle. Furthermore, the goal of the deployment project may be more research based, such as the investigation into the feasibility of a particular smart technology, or perhaps to assess its potential impact on a particular condition or client need, or indeed a combination of these.

Regardless of the overall deployment goal, be it research-based investigation, design, and development or simply the adaptation of an existing solution to a new problem context, the two core issues of *usability* in particular and *appropriateness* to a lesser degree will remain more or less the same. Both these issues are user-centered issues and given the broader context of a care setting, where typically this can range from a large health care facility such as a public hospital to a smaller residential care home or even to an individual's own home, the emphasis should be on a *person-centered* approach to deployment.

The focus of this chapter is on discussing usability, as this is seen as core to the effective design and development of smart technology solutions. The issue of what is appropriate is also considered, but to a lesser degree, as first it can be argued that a more usable care solution will likely be a more appropriate (or preferred) solution and second, the question of what is or is not appropriate as a solution is more an issue related to the specific problem context, the social and cultural norms applied, as well as the ethical and/or legal constraints imposed.

Finally, it is worth noting that this chapter is not intended to be a comprehensive, all-encompassing review of usability, but rather a representative selection of the key issues effecting the deployment and evaluation of usable, smart technologies based on real-world experience and an understanding of usability that is less oriented toward large-scale commercial applications development and more oriented toward smaller, even individual or small group-sized care settings.

It is hoped that it will provide useful insights, helpful advice, and understanding of this very broad area. The discussion draws on experience and expertise gained from a range of application domains such as mobile applications and game development, to ambient-assisted living environments. These are presented as real-world case studies. It encompasses the collective experience of both researchers and developers from across several different disciplines including commercial software and computer game development, electronic engineering, product engineering, physio-therapy, and social care. Such a cross-disciplinary approach is in keeping with best practice for usability.

The discussion first addresses usability, particularly in relation to the broader field of human–computer interaction (HCI). It asks the very probing question of what makes a smart technology usable. It then considers the approach taken to deploying a smart technology in a care setting and in particular how the usability of this deployment might be evaluated. There then follows a brief outline and consideration of four real-world case studies, where various smart technologies, devices, and sensors were deployed across different projects ranging from small-scale rapid, pilot deployments, to full-scale 3D game prototypes, to a large-scale, long-term development of an ambient-assisted living environment. Following these case studies, key lessons learned and recommendations are considered and presented.

3.2   UNDERSTANDING USABILITY

From outset, it is important to state that this discussion is not intended to be, nor can it be taken as a complete, comprehensive review of usability. Quite simply, even when considered in isolation from the much broader "parent" field of HCI, usability is still too big a subject to fully contend within such a discussion. It would take at least a full book to do so and there are many such books written on the subject! Rather, the intention is to highlight key aspects of usability such as specific definitions (again, there are many), metrics, and typical evaluation and testing methods, particularly in terms of their importance to smart technology deployment as highlighted in the example smart technology case studies presented.

Moreover, relatively recent developments informing the understanding of usability come from the domain of web-based and, increasingly, mobile usability, especially the explosive growth areas of smartphone technologies, software apps, wearable devices, and sensors that are being made available. Alongside this, we should be cognizant of the fact that a lot of what is out there about usability comes from a commercially oriented world view, where usable devices, applications, and services are seen as being more attractive to customers and hence these will provide an all-important competitive edge. In short, more usable means more profitable.

There is nothing wrong with such a view per say. It is important however that this more commercial (mass market) view of mainstream usability be balanced against a more altruistic view, more appropriate to care settings. Furthermore, commercial views of usability tend to focus, perhaps too much, on a narrower subset of attributes relating to issues such as "ease of use," "user friendliness," and "simplicity" that are generally measured against a limited set of tasks, interaction contexts, and device capabilities, typically performed and hence assessed by generally younger, more technically competent "ideal" users. While these are indeed a valid subset of attributes pertaining to what constitutes a usable app or device, there are many more such attributes relevant to what makes a technology usable, especially to a broader range of users, with varying needs, capabilities, and indeed limitations.

Finally, it is also important to note that achieving a truly usable and also an appropriate smart technology deployment, in a care setting in particular, will require the consideration of aspects not normally associated with commercial usability such as the particular special needs of a small set of users, even down to a particular individual. Such special usability needs often stem from challenges related to a specific health issue or condition. Addressing such needs will also likely require consideration of broader interaction design issues pertaining to more complex or even subtle human factors, sensory and cognitive impairments, and can also include emotional and behavioral aspects. As such, it is therefore generally advisable, as a deployment strategy, to employ the widest possible (or available) range of opinions and expertise when considering usability in this context.

3.2.1 HUMAN COMPUTER INTERACTION

In HCI, or more historically, "man–machine interaction" (MMI), "man" is of course gender plural, which traces its roots back first to the design and development of basic, everyday tools and artifacts and second to the mechanization and automation associated with human operation of increasingly complex mechanical devices and systems. Good examples might be the development of the hand drill and the development of the automobile. Alongside such developments was the emergence of domestic appliances, hence leading to mass-market consumer products, in particular electromechanical, electrical, and later electronic consumer goods. These areas were being progressed from both technological and human factors advancements, generally being pioneered from within advanced fields of technology and science such as aeronautics and later even space programs.

Hence the "timeline" of progression in terms of human factors and interaction design can be said to range anywhere form hundreds of years, if we consider the most basic tools and artifacts used for the most ordinary of daily tasks, through to at least decades for the more complex machines and everyday consumer goods. There are two important points relevant to technology deployment to bear in mind in this. First, MMI is an old "science" and second, many of the classic usability issues (particularly human factors and ergonomics issues) associated with everyday living, which are very much part and parcel of supporting real people in care settings, including within their own home, are not particularly new either.

The "modern" computer science field of HCI is somewhere between 50 and at most 75 years old. Curiously enough, the subdiscipline of "usability" is even younger, perhaps not more than 30 years old. One of the pioneers of HCI and usability was Prof. Brian Shackel, who from the late 1980s onward pioneered both the framework for considering usability, the definition of usability, and what factors contribute toward enhancing usable and effective technology design, development, and deployment (Shackel 1986).

It may come as a surprise to realize that there is no single *de facto* definition of what exactly constitutes HCI! The ACM's Special Interest Group on HCI (SIGCHI) developed a useful, "classic" definition, which was aimed mainly at educating undergraduates interested in the field, in the early 1990s: "Human–computer interaction is a discipline concerned with the design, evaluation and implementation of interactive computing systems for human use and with the study of major phenomena surrounding them" (SIGCHI 1992).

So, even with this quite basic definition, we have a focus on people and usability (i.e., "for human use"), which also puts evaluation (i.e., usability evaluation or testing) ahead of implementation! Also the, "study of the major phenomena surrounding them" is an interesting (if somewhat vague) reference to much of what forms the core issues within real deployment.

An excellent discussion on the history and more importantly the progress of this broad field, presented by Dr. John Carroll, clearly highlights this. To quote Dr. Carroll:

> Indeed, it no longer makes sense to regard HCI as a specialty of computer science; HCI has grown to be broader, larger and much more diverse than computer science itself. HCI expanded from its initial focus on individual and generic user behavior to include social and organizational computing, accessibility for the elderly, the cognitively and physically impaired, and for all people, and for the widest possible spectrum of human experiences and activities. It expanded from desktop office applications to include games, learning and education, commerce, health and medical applications, emergency planning and response, and systems to support collaboration and community. (Carroll 2014)

Clearly, the understanding of HCI and its application to real-world technology deployment has moved on considerably. In particular, such technology deployment now demands more than ever a multidisciplinary approach. The core disciplines of computer science (especially in relation to this discussion commercial software development and game development), human factors and ergonomics, as well as cognitive psychology remain, but these are being increasingly supported and supplemented by areas as diverse as medical and social care, education, business, engineering, and architecture. This list is ever increasing as the challenges being embraced continue to diversify. Indeed, even in the four case studies discussed in this chapter, such a range of disciplines is clearly apparent.

3.2.2 Usability Definitions and Metrics

There is a tendency when addressing usability is to simply present a relatively straightforward definition (or perhaps two or more) as if first, this would definitively frame a complete understanding of this issue and second, such definition(s) and any associated usability attributes or metrics will be directly applicable and unambiguous in every deployment situation. While there is no doubt that definitions, in particular standardized definitions employing clear criteria and metrics do help, for example, when scoping and parameterizing detailed design decisions and also later for evaluating and testing, these may not actually provide sufficient support particularly where the deployment of a smart technology solution involves a high degree of innovation or it attempts to address very specific, special needs. Furthermore, definitions should not ever really be seen as being all encompassing or sufficient. Asking "what is usability?" is not necessarily the same as asking "what makes something usable?" or "how do we determine that something (e.g., a smart technology) achieves a satisfactory level of usability?"

As mentioned in the beginning of Section 3.2, it is important to bear in mind that a lot of what is defined and presented on usability has emerged from and is directed toward the design, development, and testing of commercial, technology-based products and services. Thoughtful consideration and careful judgment is needed to interpret and apply usability concepts that are oriented more toward, for example, business productivity and efficiency, rather than toward the needs of a typical care setting. Also, we need to be careful specifying "more usable" as a design goal, where the real design challenges may actually relate to issues such as a persons' particular human factors needs and/or a challenging cognitive condition.

For example, what happens to be a concise, neat, and tidy usability definition or criterion in the face of real people ("users") in a day-care setting with moderate-to-severe dementia? Usability attributes relating to, let us just say "familiarity" or "affordance" or "recall," may not even be applicable or indeed appropriate! Getting back to defining usability, let us consider one such standard definition. The International Standards Organization (ISO) has done much work in this area and one definition often cited (there are many ISO standards that address different aspects of usability in different contexts) is ISO 9241: "The extent to which a product can be used by specified users to achieve specified goals with effectiveness, efficiency, and satisfaction in a specified context of use" (ISO 1998).

This definition is expanded, and made more comprehensive, by the inclusion and detailed discussion of various characteristics, which must be met by a product, for the "users of a product," such as being *effective, efficient, engaging, error tolerant, and easy to learn* (ISO 1998). Again, note the overt commercial "product" orientation in the definition. This international standard definition is based on the seminal work carried out by such usability experts as Dr. Jacob Nielsen (Nielsen 1993), who discussed more specifically the design and development of usable *interactive software* products and services and, more recently, again by Dr. Nielsen and his expert colleagues in the Nielsen Norman Group (Nielsen 2012), who considered usability more in terms of *interactive website* design.

Underpinning this work is some of the earliest works published on usability by Prof. Brian Shackel where he defined a "general framework for usability embracing the four principal components of any interaction: user, task, system and environment" (Shackel 1986). Now some 30 years on, though technologies significantly advanced, these four principal components of interaction, and hence of usability, are still directly applicable. Shackel further stated, "Good design for usability depends on achieving successful harmony in the dynamic interplay between these four components" (Shackel 1986). In other words, it is essential to consider *each* component as broadly as possible: user, task, system, and environment and in *equal* measure of importance, when considering the usability of any interactive technology. Another important point to notice is that Shackel highlighted the "dynamic" nature of usability (and in particular in terms of its assessment). Usability is fundamentally a dynamic, not a static concept. It has to be modeled, simulated, or better still, carefully observed in a "live" and realistic setting. Shackel also proposed four fundamental "operational criteria" or metrics to measure usability: *effectiveness, learnability, flexibility,* and *attitude* (Shackel 1986). He discussed how each of these criteria are then further defined (specific to each interaction or deployment context) to consider task performance (e.g., speed and/or rate of errors), across a range of users (the wider and more representative the better) and also (differing) usage environments, given sufficient training and ongoing support, within definable and acceptable levels of "human cost." These should include human factors and ergonomic issues such as fatigue, personal effort, and even subjective issues like frustration! These cornerstone components (user, task, system, and environment) and the associated four key metrics (or at least some close derivatives) have been incorporated into many formal usability definitions.

From direct experience, I can attest to the effectiveness of carefully applying these core usability components, particularly in relation to requirements gathering and early-stage prototyping, even within a technology deployment domain as complex as field-based mobile learning (Ryan and Finn 2005). Such a scenario-based approach incorporating realistic and verified user profiles, task models, context models (dealing with differing "in field" interaction environments), and specific test scenarios is still, I believe, highly relevant particularly in care settings.

Prof. Alan Dix (Dix et al. 2004) further refined Shackel's model of usability and presented a set of abstract design principles that when applied to the design (primarily) and evaluation (subsequently) of an interactive software product would enhance its usability. These design principles are expressed within a coherent catalog, which defines three broad categories: *learnability, flexibility*, and *robustness*. This is of particular importance, as these broad categories are, in my opinion, applicable and appropriate in *all* technology deployment situations. Users, better understood as clients or even actual people (persons if you prefer), first need to *learn* (and relearn) how to use a new technology (likely also, in a new context), then they require sufficient *flexibility* in how it can either adapt or at least be adapted to their particular situation and needs, and then finally they need to experience the technology as sufficiently *robust* to continue to use it including changing their own behavior, work practice, or even living patterns!

It is however important to remember at this point that no matter what definitions, models, or attributes of usability are being considered, these can never provide the complete picture. From the previous discussion on the broader field of HCI (within which remember usability fits), many of the "major phenomena" referred to within the definition of HCI, and indeed the ever-increasing scope of HCI, which is now, "more diverse than (just) computer science" essentially further complicate the actual usability issues presenting. This is especially the case where there are chronic conditions or even acute medical needs, or special care requirements presenting.

As a result, it is likely that what is often considered or attributed to a usability issue is more likely a deeper HCI issue related to some combination of human factors, ergonomics, accessibility, cognitive psychology, interaction modality, or particular sensory or cognitive capabilities and limitations. Yes, it is a long list, perhaps not even a complete list! Where such "usability issues" are either the primary focus of the deployment or are secondary but nonetheless important, it is not only advisable to have access to more specialist HCI expertise, but will likely be entirely necessary. Cognitive psychology, in particular dealing with human performance and error, is perhaps one of the most challenging areas faced. Although HCI and usability practitioners and hopefully also software developers will have at least an appreciation of the "human information processor," which is based largely on the pioneering work of Card et al. (1983), even this will likely not be entirely sufficient.

3.2.3   What Makes a Smart Technology Usable?

Strange as it may at first seem, merely referencing and applying usability definitions and various usability criteria, metrics, or heuristics during both design and evaluation does not guarantee a truly usable solution to a real-world problem such as a smart technology deployment in a care setting. The design, development, and deployment processes that deliver the solution need also to be structured and operated with usability as the central goal.

About 25 years ago, one of my own computer science lecturers, Prof. Tony Moynihan, challenged us as undergraduates to consider the future of technology, in particular how people might interact with such technology, looking ahead by about 30 years. We all agreed that such future technology would be amazingly powerful, small, portable, and cheap and that people would be able to interact with computers in very natural ways such as speech. I remember Prof. Moynihan pausing to reflect and then giving us this wise, and as it has turned out from my own experience at least, correct insight, "Who really knows for sure how technology will have developed a generation from now? But there are two things I'm quite sure about: first, hard problems will still be hard, and second, people will change and adapt very slowly."

Yes, smart technologies are part and parcel of the amazing advances in science and technology witnessed in the last quarter of a century or more. But making them really useful and indeed more available and accessible, especially to a diverse range of nontechnical users, particularly within a health care setting, remains a hard problem to solve, one in which user behaviors and attitudes can be very slow to affect.

Surprisingly, understanding such a "hard problem" as making interactive computer technology usable was at the forefront of the discussion as far back as 30 years ago! In a seminal paper, "Designing for Usability: Key Principles and What Designers Think," John Gould and Clayton Lewis addressed this very issue:

> Three principles of system design which we believe must be followed to produce a useful and easy to use computer system. These principles are: early and continual focus on users; empirical measurement of usage; and iterative design whereby the system (simulated, prototype, and real) is modified, tested, modified again, tested again, and the cycle is repeated again and again. (Gould and Lewis 1985)

The potentially open-ended and iterative nature of modification and test cycles suggested went very much against the grain of the establish norms of the linear, "waterfall" process models that were considered best practice in that era. Thus, Shackel elaborated and expanded on these fundamental features of design for usability:

1. *User-centered design*: Focus from the start on users and tasks.
2. *Participative design*: With users as members of the design team.
3. *Experimental design:* With formal user tests of usability in pilot trials, simulations, and full prototype evaluations.
4. *Iterative design*: Design, test, and measure, and redesign as a regular cycle until results satisfy the usability specification.
5. *User supportive design*: Considering training, manuals, quick reference cards, online help, and so on (Shackel 1986).

Added to the potential project management nightmare of open-ended iterations, we now had the suggestion that real-end users participate directly as members of the design team! Even 20 plus years after this, usability experts such as Jeffrey Rubin (Rubin and Chisnell 2008) provide pretty much the same advice: *early focus on users and task, evaluate and measure usage and iterative design and testing.* Also, as part of an excellent discussion on what makes something usable, they pose two very insightful questions: "What do we mean by usable?," within a given problem context and in particular, "what makes something *less* usable?" It came as a real surprise to me to seriously consider, as part of design and development, what factors in my own understanding, approach, and processes might actually be *detrimental* to usability! Rubin outlines five very compelling and challenging reasons of why (software) products are hard to use, all of which I can, from experience, identify with and fully endorse:

1. Development focuses on the machine or system.
2. Target audiences expand and adapt.
3. Designing usable products is difficult.
4. Team specialists do not always work in integrated ways with separate system components being deployed independently.
5. Design and implementation do not always match (Rubin and Chisnell 2008).

It is interesting to note that more than 20 years on, usability experts explicitly highlight that designing usable products is a hard thing to do! Much of the additional wisdom reflected in this list is also apparent in the case studies discussed in this chapter.

3.3 APPROACH TO SMART TECHNOLOGY DEPLOYMENT

This section does not seek to provide a definitive answer as to how to approach the deployment of smart technology in a given care setting. Rather, it offers advice based on real-world experience from a range of different deployment projects. The discussion is also informed by what is considered to be best practice. The intention is to help enlighten the consideration of deployment strategy, primarily from the perspective of usability. It is important to bear in mind that every case is different and requires specific and careful consideration of the local context, circumstances, and needs as well as the likely more immediate issues associated with availability of necessary resources such as funding, people, technical support, and the constraints imposed by project timescales, and so on. Furthermore, the emphasis of the discussion here is on highlighting key issues effecting users and usability. There are other equally important priorities associated with deployment, such as in relation to research and development aims, which though addressed are not discussed in depth.

3.3.1 WHAT IS THE GOAL FOR DEPLOYING THE SMART TECHNOLOGY?

As was stated in the introduction, *well-being enhancement* should be the ultimate goal of deploying any smart technology in a care setting. In short, whatever smart technology is deployed for whatever reason, overall it should impact positively on the well-being of the individuals affected. It was also suggested these enhancements need to be specific, tangible, and also measurable. However, such well-being enhancements are often secondary to a more specific primary goal, typically associated with a given research aim. Hence, well-being enhancements are usually achieved indirectly as a result. Furthermore, different care needs and situations necessitate different models of care being employed. For example, there might be a carer or team of carers, and perhaps family members along with professionals, tending to a client or group of clients. There might just be a single person taking care of him or herself either independently or perhaps with a degree of dependence. In all such care scenarios, it is important that both primary and secondary goals are understood, agreed, and achieved to some degree at least.

The primary goal can also relate to helping or supporting an individual or group to *change* some aspect of their own *behavior* or life-style pattern, for example, when they might go to bed or by being more engaged and active. Achieving such a primary goal as behavior or life-style change should also by definition produce some of the specific well-being enhancements sought, such as perhaps losing weight, becoming more dexterous, or being less anxious or more socially engaged, or whatever. The establishment of a clear cause and effect link between such primary and secondary goals may form a significant part of the research work required and may indeed be part of the investigation being undertaken.

Alongside these primary and secondary goals, there is also (and always) the need to provide both a *usable* and also an *appropriate* solution. At least satisfactory levels of usability are required and hopefully for the widest range of users, for example, clients, carers, family members, health professional and managers, and so on. Each group will likely have differing needs and priorities. There is also the aim(s) associated with the deployment project itself, be it mainly a research-based investigation, product design, development and testing, a feasibility study, or perhaps just providing a workable solution to a particular need in a specific case. There will most likely be a combination of these.

Finally, different constituents involved or related to the project may have differing interpretations of the goal(s) and priorities set. Typically, funding or regulatory interests will be focused on cost, effectiveness, and deployment schedules. Advocacy groups will more likely tend toward provision of what they would see as appropriate solutions. Overall, though the overriding goal(s) should reflect more the interests and priorities of the ultimate end users, particularly their well-being as well as their ability to effectively use, and hopefully, their favorable disposition toward the smart technology deployed.

3.3.2 BEST PRACTICE AND BEST ADVICE

The following five issues represent a *prioritized* list of what are seen as the *must do* (or at least must address) essentials when deploying smart technologies in a given care setting. These are intended to bring together both the practical experience (such as that represented by the case studies outlined) and an assimilation of what is considered best advice, representing a range of expert opinions, much of which has been previously discussed around HCI and usability in general. These must do items should provide an effective starting checklist to ensure that the core issues relating to the usability and suitability of the technology deployment are considered. The presentation is kept short and to the point so as not to distract with too much detail, as the intention is mainly to reassure in terms of the overall approach adapted.

Again it is important to reemphasize that each particular deployment differs in terms of the context, setting, health, social, cultural, and regulatory challenges present. Also, the range of expertise, ability, experience, and even the availability of the researchers, developers, and care practitioners (be they carers, managers, medical specialist, etc.) involved will vary.

3.3.2.1 Be User Centered

This is the most important issue relevant to achieving a usable solution in all cases. It is pretty much given, maybe even a truism that you must, from the outset and throughout the project, be entirely user centered. But we need to be careful of how we address the "spirit of the law" and not just the "letter of the law" particularly when it comes to care settings. We can tick all the boxes in terms of conducting user analysis, profiling, and even demographic research. We can produce comprehensive user personas and verify the accuracy and dependability of these through the most expert user proxies available.

This will still leave at least one essential aspect outstanding. Commercial usability and UCD techniques tend to focus design on "everybody," even where specific personas are deployed to ensure that they adequately represent particular market segments or niches. Within care settings, however, we need to remember that being truly *person centered* is the key. We must *design for* and indeed *with someone*, ensuing that each profile (or user classification) is adequately and accurately instantiated and represented directly and specifically by actual persons. It sounds simple and indeed for smaller scale projects it may be a relatively easy thing to ensure. Obviously, the larger the scale of the project deployment the harder it becomes to maintain this focus, but it is all the more necessary. Being truly user centered also means having full user *participation* in the design, development, testing, and deployment. Within a care context, this can and often does present additional challenges, for example, relating perhaps to the advanced age or reduced cognitive ability of individuals, but these issues can be mitigated especially through the use of expert user proxies who can reliably represent the opinion, response, and abilities of a specific user group.

3.3.2.2 Iterative and Agile Development

From the previous section, the case has been well made and indeed remade for design, development, and deployment projects to operate *iteratively*. This is particularly important the more the project is oriented toward fundamental research, or innovation or deploying cutting edge technologies. You would not get it right first time and often you do not intend to. A lot of interesting findings can still be produced even where deployment is not (at least initially) particularly successful from a usability point of view. We need to be careful however not to view project iterations as second, third, or subsequent chances to "get it right" or "fix user issues." Nor should we fall into the trap of thinking that an iterative approach means we are embarking on an open-ended project management approach! Iterative development does not mean we "go round in circles" endlessly, not least because very few if any projects have unlimited resources available!

Iterative design and development has really come of age with the application of modern software project and process management tools, which provide for *agile* development. One such toolset, which has been very effectively applied in some of the case studies, is Scrumwise (2015), which provides a collaborative environment for development team members to effectively manage all aspects of a project throughout the full design, development, and deployment life cycle. Not only does such an agile approach provide for complete flexibility, but it is also at its core customer (or care client) focused, particularly in terms of constantly striving to deliver functional components that demonstrably match explicitly stated and well-managed requirements. Both software developers and other project specialists including expert user proxies and even end users can be involved in managing the identification, prioritization, and delivery of these requirements, collaboratively.

Finally, agile iterative development can scale from very small, even single developer-based rapid prototyping, providing workable release cycles of less than a few weeks, all the way up to large-scale, multideveloper projects that may run over release cycles of several months or more. Such larger projects tend to define major delivery stages such as a proof-of-concept prototype, followed as necessary by increasingly sophisticated and more functional software releases as suits the deployment. Typically between two and three such incremental releases will yield a completed solution.

3.3.2.3 Multidisciplinary Teams

As has been previously highlighted, the scope of HCI has expanded well beyond computer science. There are very few areas, if any, where smart technologies, for example, smartphone apps augmented by environmental, context, and even body sensors, are not being deployed and as such, domain experts within the particular specialist deployment field should (obviously) be a part of the project team. Indeed for most projects in areas related to health care, the lead team members will likely be those with expertise in the particular type of care need or the condition being addressed, and not those with computer science, software development, or HCI-related skills.

Even from the small set of case studies considered, teams typically included people with skills and expertise in disciplines ranging from software development, project management, HCI and usability, electronic engineering, multimedia (including 3D graphics, audio, and music production), game development, to physiotherapy, occupational therapy, psychology, health care, social care, architecture, and education. This list is not actually exhaustive nor is it in any particular priority order! Neither does it necessarily include the additional experts, advocates, and experienced end users or user groups necessary, from within the particular care area being addressed.

Trained and experienced professionals from such fields, particularly those with expertise in research and development, will not only understand the need for, but also be well equipped to deal with the complex issues stemming from project management, group dynamics, interpersonal communications, decision making, and leadership. A key point worth highlighting is that central to all of these "group dynamics" is the importance of maintaining a clear focus on and representation of users and their specific needs and issues, particularly those related to usability. This focus also will help foster and encourage clear communication, creativity, innovation, and even fun!

3.3.2.4 Universal Design

Universal Design (UD) was first proposed and defined in the 1990s by Ron Mace and his design group, based at North Carolina State University in the USA as, "the design of products and environments to be usable by all people, to the greatest extent possible, without the need for adaptation or specialized design" (CUD 2008).

UD has since been widely adopted in many countries and is now accepted as a model of best practice design particularly in relation to addressing the needs of "nonstandard" users. Indeed, the definition has been further refined, and in certain countries, even adopted into legislation. For example, the Centre of Excellence in Universal Design in Ireland has amended the definition (within a National Disabilities Act) thus: "...accessed, understood and used to the greatest extent possible... by any persons of any age or size or having any particular physical, sensory, mental health or intellectual ability or disability" (CEUD 2015a). This further refines usability (i.e., "to be usable by") so as to include "accessed, understood and used" and furthermore it explicitly states that this applies to the widest possible range of users, indeed *any* persons, of any age, size, physical, sensory, mental, or intellectual ability or disability!

So where such definitions and requirements are stated under legislation in particular, they cannot be ignored. Indeed, it should be clear to anyone considering deploying any kind of smart technology in a health care or care setting, such requirements should not ever be ignored or overlooked. In fact, it is likely the case that there is much more to be gained by proactively embracing a design challenge posed by UD. There is of course the fairly obvious reason that because the deployment is aimed at some kind of care or health care need, it will very likely and in reality almost definitely includes people for whom usage of technology would normally require a degree of specialist adaptation. But there is a less obvious reason, which relates to the innovation potential of actually catering to the special needs of nonstandard users from the outset. The spin-off benefits generally tend to favor a wider spectrum of user groups if not indeed all user groups.

Explicitly embracing UD as part of the deployment strategy also means that designers and developers can access an array of design principles, development methodologies, resources, and toolsets being created and made available around this whole area (CEUD 2015b). Indeed, UD can also serve to lift the taboo associated with many disabilities, emotional, and behavioral issues and even mental health conditions and disorders. It can also help unify a design as being both *usable* and more *appropriate* for users within both the "normal" and "abnormal" range of abilities or disabilities. Finally, it carries a high degree of altruistic as well as social-capital potential.

3.3.3 Considerations for Evaluating and Testing the Deployment

Any deployment of a smart technology in a care setting is *de facto* a usability study of sorts, even almost unimaginably, if little or no formal usability definitions, metrics, or methods are considered or if there is no usability expert or specialist explicitly assigned to either the research and/or development team(s). It is considered as unimaginable to explicitly ignore or even detach usability completely from a research project, as a key part of the research method will most likely relate to the evaluation of the deployment. Likewise for development projects, there will at least be some form of basic user acceptance testing, as part of each, successive iteration. Where a project combines research and development (as one or, for larger projects, two or more teams), then usability should be central to both.

There are an array of techniques available addressing usability evaluation and testing both from a HCI research methods perspective (Cairns and Cox 2008) and from the IT and consumer products perspective (Karwowski et al. 2011). There is also any amount of resources, materials, and even toolsets covering usability evaluation and testing available online such as the noncommercial, US Government resources available (Usability 2015) as well as services offered by commercially oriented usability consultants such as Nielsen Norman Group (NNG 2015).

In any case, the scale and scope of what can and should be attempted is largely dependent on the budget available both in terms of time and money to the project as a whole. It is likely that deployment trials, be they explicitly research-based trials, or development testing, or indeed a combination of both, will amount to a significant proportion of the overall project work undertaken. It is also advisable and desirable that at the earliest stages of the project, direct user input is sought, likely in the form of brainstorming sessions or focus group discussions or the like. There may also be the need to access the opinions of expert user proxies in the case where direct user input is either not available or perhaps not appropriate. Furthermore, there are no real hard and fast rules to which evaluation approach might best suit. Indeed, it is more likely that a mixture of HCI research methods, usability evaluation, or testing techniques, and increasingly user experience (UX) techniques as outlined by Albert and Tullis (2013) may be applied. It should also be remembered that common sense needs to prevail as well as practical constraints imposed by limited resource availability.

Finally, it is most important that we maintain a high degree of contact, involvement, and empathy with the widest range of users, or at least proxy representatives, as possible. An empathetic approach is at the core of how care and health services need to be delivered. Being person, patient, or client centered should not just be an aspiration but should actually be the norm in care and therefore also in the evaluation of usability within that care.

3.4   SMART TECHNOLOGY DEPLOYMENT CASE STUDIES

The following four case studies are presented in brief outline within this section. Each has been chosen so as to highlight and exemplify some of key aspects in the discussion on smart technology deployment and usability. More detailed information can be found by following the references given. The first three projects are representative of ongoing research and development within the Well-being Enhancing Technology (WET) research group based at Satakunta University of Applied Sciences (SAMK), Pori, Finland, in collaboration with Dundalk Institute of Technology (DkIT), Ireland. All three projects attempt to explore the impact of serious game design and gamification on well-being enhancement and also form the basis of the discussion presented in Chapter 13 of this book.

This was a new departure for DkIT, as although the Institute has within its Department of Computing and Mathematics extensive experience in the field of advanced 3D game development, applying this to real-world problems within a care or therapeutic setting was a step into unknown territory! Much has been made of the potential for serious games and gamification in real-world problems, including within health care by leading experts such as McGonigal (2011).

The fourth case study is based on work done within DkIT involving its Netwell Research Centre and the spin-off commercialization group, CASALA. It is also important to note that this fourth case study, *Great Northern Haven*, which is considered here from a usability perspective, also forms the basis of the discussion presented in Chapter 6 of this book.

In terms of scale, the first two case studies *Activation and Rehabilitation Games for People with Special Needs* and *Cognitive Mobile Games for Older Adults* represent *small-scale* smart technology deployments, both in terms of the small size and relative simplicity of the software game apps produced. The third case study, *Well-being Enhancing 3D Games*, is more typical of a *medium-scale* deployment, particularly in relation to the much larger size, scope and complexity of the 3D game prototypes produced. The fourth case study, *Great Northern Haven*, typifies a *large-scale* project insofar as it was ran over a number of years, involved extensive user needs and requirements gathering, and the design and construction of purpose built, customized smart homes in a full-scale residential complex that has been fitted out with a range of smart technologies, devices, and sensors and is also being evaluated, as a living lab, on an ongoing and continuous basis.

3.4.1 Activation and Rehabilitation Games for People with Special Needs

In essence, this study combined exercise and gameplay on a smartphone device as a means of both activation and rehabilitation for people with a combination of special physical, cognitive, and developmental needs. A summary of this work is provided by Koivisto et al. (2015) and it is further discussed in Chapter 13 of this book. Simple, intuitive, custom-made smartphone game apps were designed and developed that challenged users around relatively straightforward physical exercises such as holding and moving the smartphone to articulate position. This was extended by cleverly using the smartphone's motion sensor, while the phone was simply mounted into a modified exercise balance board, which can support a user standing on it or using their feet while sitting, or even their hands while kneeling. The smartphone was wirelessly networked to PC, which connected to a large flat screen display or TV. Games were designed around a simple ball and bat Ping-Pong theme. These were trialed in a range of care settings, with both older and younger adults with various special needs.

A further extension to this concept saw the development of a modified painting type app, also smartphone or tablet PC based. The app can be networked to another smartphone allowing one user to paint and the other to either paint along with or assist with the selection of colors, paint tools, and so on. A networked PC facilitated the projection of the image, which was also overlaid with various reference artworks and backdrops. As part of the KOLMIO Project (Sirkka and Koivisto 2015), this combination of "art and technology" was deployed and trialed at the Kankaanpää Rehabilitation Centre in Finland involving clients in rehabilitation typically following limb injury or stroke, as well as clients with more severe mobility impairments.

Care was taken when designing the game apps for mobile devices, that is, smartphone or small tablet PC. Human factors and usability issues particular to small screen mobile devices have been well researched and documented by experts such as Nielsen and Budiu (2013), and every effort was made to address these. Also, from a design point of view, the emphasis was on providing ultrasimplistic, intuitive features.

At its most basic, these initial game trials, especially with the paint game, involved a wide range of users with varying degrees of physical mobility or disability issues simply articulating forward and back, and left and right (X and Y) movement. In Kankaanpää, in particular, the trial began with able-bodied players (such as the researchers and app developers) playing the game as a demo in an open communal space within the center. People began to stop and observe and then wanted to have a go. Engagement tended to progress from people with minor incapacitation, to those with more severe disablement, and finally to a client with quite profound disability who was using a sophisticated electric wheelchair operated by a joystick for mobility.

From a usability point of view, these games were simple and intuitive, self-learning, and socially very engaging. But even for such simple artifacts, they did present some challenges and confusion especially around multiplayer interaction. They showed a remarkable degree of flexibility in how users could articulate and cooperate, and also proved to be very robust. In short, something worked to a user's satisfaction in almost every situation! There was also a high degree of user innovation shown, for example, two players created pictures resembling a simple racing track and then proceeded to play an improvised go-kart type racing game by simply trying to "paint" (or drive) while trying to keep their "brush" (or kart) between the track edges. The racing element of the game came about by each trying to be the first to get to the finishing line they had previously drawn! The trial also supported deep empathy between developers, researchers, and users, being hands on, cooperative, and socially engaging across a very wide range of user abilities. They also provided useful design inspiration such as providing for instant color printouts of peoples "art" and the possibility of sharing these via social media.

3.4.2 Cognitive Mobile Games for Older Adults

"GaMeR" (Games to Support Memory Rehabilitation) was designed and developed to explore the link between game play and memory function in older adults with dementia. Two combined game apps were developed for tablet PC, the first involving a modified, interactive version of the standard MME memory test to assess cognitive and memory function over time and a second simple cat and mouse game where the player controlled their own mouse character by tilting the tablet PC device to move their mouse at varying speeds to collect pieces of cheese and also avoid being caught by the chasing cats! This second game assessed dexterity, motor function, hand–eye coordination, and reaction. It also provided stimulation and engagement.

A full discussion and analysis of the research trials conducted in Finland are given by Merilampi et al. (2014). These were conducted over a three-month period as part of the GaMeR—research project in a War Veteran's Nursing Home and Rehab Centre run by West-Finland's Diaconese Institution in Pori Finland using a formal research method involving a trail group and a control group. An important point to note here is that each player's scores were recorded and logged anonymously by the game on a secure server for later analysis. Careful attention was paid to the design of the games through initial focus group meetings, which addressed the usability of the game design with UX interviews being conducted at the end of the trail. The initial design meetings focused on important human factors and cognitive aspects of the game interface, simplifying functionality and enhancing comfort especially as these games were targeted toward older adults with mild-to-moderate age-related dementia.

As members of the WET research group had previously participated in comparison trials studying the UX of a tablet PC-based maths game for school children (Kiili et al. 2014), the benefits of comparison trials across different nationalities, cultures, and user contexts were well understood. Also, these comparison studies employed a mixture of different user evaluation methodologies. Based on this experience, the opportunity to conduct a similar cross-national comparison of GaMeR was explored. Although an identical (or at least very similar) setting in terms of participants' age, care environment, care model, and condition severity was sought, none could be found. However, a reasonably similar setting was identified within a day care setting for older adults (as opposed to a full-time nursing home setting in Finland) but for adults with moderate-to-severe dementia. A more detailed discussion on this study is provided by Finn et al. (2015). Due to logistical issues, the impact of significant restructuring of the care services and an increase in the workload carried through the care setting, progress of the trial was delayed and although the trial period was extended, the study was unable to be completed as planned. These realities are very much part of the risks and resource limitations associated with real-world live trials, especially in the voluntary care sector. Furthermore, due to the moderate-to-severe nature of the dementia presenting in participants, game interactions were more time consuming and difficult than in the Finnish trial. Clients were far more dependent on carers' assistance to successfully progress. This was further complicated by the changeover and introduction of new staff during the trial.

Despite all of these difficulties and setbacks, from a usability point of view, this comparison trial proved fruitful. Some key issues were identified and deep insights were made. For example, despite clients having to relearn game interactions, there was a high level of positive engagement and participation that produced a noticeable, often momentary benefit arising from the cognitive, physical, and emotional involvement in the game tasks. Due to the relative sparseness of the dataset gathered, as most clients did not attend and hence interact with the games every day, it was not possible to reliably track progress in terms of memory function.

However, through direct observation and focus group discussions it was clear that there was in general quite a positive experience of engagement in game play, where clients could achieve at least some success. This provided a positive alternative to the typically more negative, pen and paper-based standard memory testing associated with the tracking of this typically declining condition.

Finally, the extent of two-player interaction, which is between the client and their carer, had not really been factored into the design of the games as these had been built to suit the requirements of the Finnish setting. This presented some usability issues relating to flexibility and effectiveness. Had there been time and resources available to address (iteratively) the adaptation of the game models, it was felt that there was real potential in fully supporting such a "two player mode." Indeed, it was felt that such interactive games, which also provided for recording game performance and relating this to memory testing, could make very useful tools for more flexibly and effectively monitoring clients progress, as well as assisting carers in relation to tracking the progress and management of clients condition.

3.4.3   WELL-BEING ENHANCING 3D GAMES

The DkIT's Computing and Mathematics Department has been delivering a specialist degree program in Computing in Games Development over the last 10 years. As part of the final year project within this degree program, students work in teams of three to five developers, each taking on a specific specialist role, for example, game engine programming, games physics, artificial intelligence (AI), interface design, 3D modeling and animation, audio and sound design, and so on. to produce fully functional, console ready game prototypes. Teams are supervised and managed by a panel of technical experts and employ state-of-the-art tools and agile development methods. Iterative development is at the core, as is expert user proxy evaluation. Technical excellence and playability are the key requirements for the games produced.

As part of the ongoing research and development cooperation with SAMK, the decision was taken to thematically move in the direction of designing 3D games to support well-being enhancement for nonstandard users, through the application of UD principles. As such, real-world clients with specific care needs were identified. These ranged from physical conditions such as muscular dystrophy, to cognitive conditions such as dementia and even emotional and mental health issues such as anxiety and depression.

In 2013, three game prototypes were developed and evaluated in cooperation with SAMK and a Finnish client care home, Kaunummen koti, as detailed by Finn et al. (2014). These games also involved the deployment of various smart sensors such as the Leap Motion Controller (Leap 2015) to augment and even replace the standard games console controller, allowing for hands-free interactions and the use of immersive 3D virtual reality (VR) headsets such as the Oculus Rift (Oculus 2015). Also, sophisticated 3D game worlds were created based around client specifications and needs to facilitate explorative and cooperative gameplay, including creating 3D replicas of actual arts and crafts artifacts created by clients within their own care setting. Additional devices were designed, such as a "mood board," that provided enhanced color lighting effects during game interactions. Following initial client evaluations and with the assistance of additional expert evaluation, one team have successfully embarked on the commercialization and setting up their own game development company, "Mega Future Games," and are progressing to full-scale user trials of their game (Mega 2015).

In 2014, the well-being enhancement theme was repeated and further enhanced. One team, "Red Ember Games," have developed "Haven," which is a fully immersive 3D environment (again employing the Oculus Rift VR headset) for children with anxiety, behavioral, and emotional disfunction. Through the support of experts in the field of psychology, accessibility, and cognitive behavioral therapy (CBT), the game prototype has been iteratively developed as a serious game to help develop and assess children's emotional and social well-being as part of CBT sessions. The game uses a sophisticated AI-based character to encourage empathetic response and was showcased by the team at the Irish Game Based Learning Conference where it won the award for best student project (IGBL 2015).

A key feature of these game development and deployment projects has been the direct involvement of a wide range of experts from across disciplines including accessibility, social care, physiotherapy, CBT, and occupational therapy providing user proxy evaluation on all design iterations from basic concepts, through paper prototypes, to proof-of-concept digital prototypes all the way up to and including interim and completed game prototypes. Both usability and playability issues have been assessed and addressed at each stage. Such a game design extends basic usability issues to providing for fully immersive, coherent, and comprehendible gameplay. An important attribute of this is to achieve flow and satisfaction during gameplay across the widest range of user abilities. This has proved quite challenging in almost all situations.

Although formal deployment trials are only just being considered for these game prototypes, extensive user testing and play evaluation was conducted as part of iterative development, including "live" demo and evaluation sessions with actual end users in realistic care and therapeutic environments. Judging by the initial reactions to these sessions, there exists serious potential for such 3D game prototypes, which is being further pursued. Both well-being enhancement and UD have provided very suitable themes for game developers to work with and have elicited a very high degree of creativity, ingenuity, and even innovation in response to the challenges presented within the care settings considered.

It is worth noting that when dealing with people with special needs, great care should be taken when considering or indeed using immersive, 3D technology such as the Oculus Rift. In all cases discussed here, the game prototypes could be used in a standard (nonimmersive) screen mode. Furthermore, the same care and consideration also needs to be applied to the visual design theme(s) used for the 3D world. In one such case, part of the game world (level design) was "toned down" thematically (from a darker castle dungeon setting to a brighter, Spanish Villa type setting) based on expert user proxy feedback. It was interesting to notice a more positive engagement with this new environment from all users tested, not just more special needs users.

There is also consideration and investigation being given to providing real-time measurement of users' emotional state and response through for example the monitoring of users microfacial expressions. Indeed, other biometrically based attributes such as measuring stress through Electro Dermal Activity are also now being investigated (PIP 2015). These have the potential to greatly enhance the therapeutic benefit of such game interventions.

3.4.4 GREAT NORTHERN HAVEN

Great Northern Haven (GNH) (Figure 3.1) is a research project examining the ability of ambient assistive technologies to support older adults to age in place. More details of the sensor technologies installed in the 16 apartments are given in Chapter 6 of this book. In this section, we consider usability and design of the apartments as well as design of software applications to support the residents in GNH.

Three public bodies work as partners in this project. The DkIT has a research interest in the project through the NetwellCASALA Research Centre. The Health Service Executive provides all of Ireland's public health services, in hospitals and communities across the country. Louth County Council is the local council, and the 16 GNH apartments are part of the council's social housing stock.

A key element of the design of the apartments was that they should not feel institutional. The apartments are designed with a minimalist style with white walls and oak flooring (Figure 3.2). The furnishing of the apartments is the choice of the residents. The feeling on entering the apartments is of entering a modern, high-quality apartment rather than any form of care facility. There is also a shared meeting space, called the parlor at the entrance to the apartments. Social gatherings and iPad training classes are held in the parlor.

A second key design element of the apartments is that of adaptability. As a resident's needs change, the apartments can be easily reconfigured to make them suitable for independent living. All the apartments have two bedrooms, between 81 m^2 on the ground floor and 72 m^2 on the upper floors. The second bedroom is much smaller than the main bedroom. It is designed for visitors or a carer in the case of the resident being dependent. The partition wall between the bedrooms is nonload bearing and does not contain any services or wiring. This is to facilitate the reconfiguration of the living space if necessary. The reinforced ceilings allow the installation of hoists for highly immobile residents.

FIGURE 3.1   Great Northern Haven, Dundalk, Ireland.

FIGURE 3.2   Interior design of GNH apartments.

UD to cater for the visually impaired, deaf, wheelchair bound, and infirm was employed in the apartment design. In trying to avoid an institutional feel, the design team sometimes neglected UD. The white color palette results in poor design for the visually impaired as there is no contrast between internal doors and walls. The absence of handrails throughout the apartments does not cater for the infirm.

The biggest usability problem encountered by the residents was with the heating controls (Figure 3.3).

FIGURE 3.3   Heating panel controls GNH.

Each apartment is divided into six heating zones, which are regulated by the pictured control panels. The control panels allow a temperature differential of plus or minus 3°C around a set temperature of 21°C. Each zone's temperature can be set individually.

There is a selectable away from home setting, which lowers the set point to 16°C. Once this selection is made on any of the controls, it is applied to the whole system. The way from home selection is indicated by an image of an empty house with a person outside. There is a further symbol of a crescent moon, which is a night setting. This has been disabled.

There are two more symbols on the control. The symbol of raindrops is disabled. The frost symbol indicates anti-frost mode. This mode is automatically engaged when a window is opened. It operates by turning off the heating in that zone. If the temperature in the zone drops below 10°C, the heating is turned on to prevent further temperature drop.

Good UD would include designing for the visually impaired. The symbols on these controls are from 3 to 6 mm across, which would make them unreadable for the visually impaired. The panels are located at waist height (to assist wheelchair users) but this make reading them more difficult.

The heating in the apartments is permanently on subject to thermostatic controls. Several residents have reported opening windows to turn the heating off until the temperature falls below 10°C and leaving them open. One or two residents forgot to close windows in very cold winter conditions.

Another major issue with the heating is that it is controlled by underfloor heating. The control panels in the six zones of each apartment open and close valves, which feed floor slabs with hot water. As the whole floor slab must heat or cool for a temperature change to occur in a zone, a considerable time lag of 3 to 4 hours is typical. The disconnect between making changes at the control and the temperature actually changing has led to the following comments from residents.

Resident 7: "I'd rather have radiators than underfloor heating."
Resident 9: "That's a big minus—you can't switch off the heating."
Resident 5: "See that heating? It's driving me around the bend. I'm not kidding!"

To try and avoid similar problems with the software we produce from the data gathered at GNH, we use an iterative approach involving the residents as codesigners. We discuss with the residents what technology they need, build a prototype, and then ask them how we can make it better. One of the first requests was for an application that would allow the residents to see their energy usage. The meters for the 16 apartments are located in the parlor, behind a locked door. This meant that residents were getting estimated bills, which they considered too high and they were interested to see their actual energy usage.

We built an initial prototype and tested with the residents (Figure 3.4). Users were asked to carry out a series of tasks and comment on their experience with the application. One of the main issues that arose was the need to make the application customizable.

FIGURE 3.4 User testing with GNH residents.

Residents needed to be able to customize the font size and background color of the application. One of the residents had an eye condition, which made reading the text on the prototype background difficult.

A home security application was developed that shows whether an external door or window was left open, the stove was left on, or there was a water leak (Figure 3.5). Green indicates a safe or closed state, while red indicates that an external door or window was left open or a stove or tap was left on.

Residents have been living in GNH since June 2010. During this time, the key themes that have emerged in the design of the living space and software to support the residents are adaptability, codesign with the residents, and an iterative approach to design and development.

FIGURE 3.5 Home security application showing an external door is open and the stove is on.

3.5 CONSIDERATIONS, RECOMMENDATIONS, AND LESSONS LEARNED

There is much more that can be said and maybe indeed need to be said in relation to smart technology deployment in care settings. It is, as can hopefully be seen even from the limited scope of this discussion and the selection of case studies considered, a very broad topic. However, there are three main issues that clearly stand out and merit some reflective consideration. These are, first, the deployment context, which captures a lot of the *where, when, and why* of the deployment. Second, there are the opportunities and barriers to deployment, which generally tend to be the flip side of the same coin, issues that are a barrier from one perspective can also present a clear opportunity from another and vice versa. Third, usability and well-being enhancement that are considered together, while not actually being one and the same thing, neither does one automatically imply the other.

For all three of the issues highlighted, deployment context, opportunities and (especially) barriers to deployment, and finally usability, *accessibility* can and often is a key overriding concern. Put simply, in deployment cases where there is serious disabilities evident, particularly physical disabilities but also sensory disabilities and/or cognitive impairment(s), accessibility needs, and limitations must be clearly identified, understood, and addressed from a broader perspective (e.g., through the provision of an appropriate adapted living environment) before any meaningful improvement or impact can be reliably determined from the deployment of a smart technology solution.

Findings are not presented as either *de facto* nor as empirically demonstrated. That is not the purpose of this discussion, which is what formal research and deployment trials seek to do more rigorously. Rather, the intention here is to provide some considered and hopefully useful advice, general recommendations for consideration, and important lessons that can be learned in relation to how usability impacts on deployment.

3.5.1 DEPLOYMENT CONTEXT

One of the classic mistakes made in design particularly and therefore by extension in development and deployment as well is doing this *divorced from the context of the problem.* To whatever degree those deploying solutions are separated or detached from the problem context, be it by lack of user centeredness, or lack of familiarity or experience within the problem domain, or even lack of direct access to the domain and domain expertise, whatever the cause and there are often many such interrelated reasons, this will almost invariably impact negatively on both usability and appropriateness. So, while specific types of devices and sensors are generally applicable, for example, the radio frequency identification (RFID) tags integrated into identification (ID) badges for GaMeR, or the standard smartphone potentiometer sensors that were used for the exercise balance board game and the standard passive infrared (PIR) motion detectors and magnetic access switches employed in the GNH, each needs careful consideration and evaluation within the specific deployment context.

Furthermore, there are standard software architectures and design patterns, generic application models and user interfaces, and so on. Likewise, we can and we want to build solutions in one real-world domain and then seek to apply these in others, for example, a lifestyle and personal fitness training app being brought into a care setting, or maybe even a "find my phone" or "location tag" app being used as a tracking device for an elderly relative who may be at risk of wandering off. Deployment can be driven "top down" from policy and planning, through investment, to a project pilot into a wider role out in a particular type of care environment. But deployment can and frequently does happen "bottom up," were individuals who have a real need, as well as the ingenuity and technical knowhow, simply apply what is available to that need and just see what happens.

This may (or perhaps may not) all be fine, but it is very much part of finding real-world solutions to real-world problems in the face of limited resources and changing needs. However, in terms of conducting deployment trials, pilots, or turnkey solution provision in care settings especially, it is essential that a very proactive, hands-on and experiential approach be taken. Researchers, designers, and developers need to be as fully immersed in the deployment context as is possible and need to fully engage with and involve those who are part of this context. Knowing that a smart technology is easy to learn and use or that it is (initially at least) favorably received is important and positive, but the full deployment may involve more challenging situations, setbacks, cutbacks, and other such real-world issues. Just how much was the flexibility (or indeed inflexibility) of the solution tested and across what range of conditions?

Also how does the deployment challenge and therefore assess or at least inform in terms of the robustness, effectiveness, and efficiency of the solution, particularly when it is measured over time? Yes we can and should conduct a longitudinal study, but that might only consider the effects or long-term impact, which is not the same as attempting to engineer in longevity and successful sustained usage.

Also, every context is different, even if only appearing to be slightly different when considered in terms of broad "macro" factors such as cultural, social, structural, and policy issues. Deploying the same smart technology solution from one problem context to another as was experienced with the GaMeR case study can be like trying to force the proverbial round peg into a square hole. However, even where there is such a mismatch this can and generally does yield much that is useful for research and design evaluation even if it (partly) fails to provide a workable solution. One size generally does not fit all in care applications. But at the more "micro" level, at the level of actual client groups, and particularly at the level individual clients, subtle but vital differences become more apparent. User centered becomes client or patient centered, which in turn becomes person centered, actually designing for somebody and hopefully developing and hence deploying with that somebody. Yes there is a degree of idealism in this, but at some point, particularly in care this must happen to be truly effective. In the example of the well-being enhancing 3D games, a large degree of customization, personalization, and context sensitivity was achieved, even at such a micro level of individualization. This deployment was perhaps even ideally contextualized. What was most encouraging was that at the evaluation stage, the games and the technological setup provided proved to be usable and playable across the full range of clients identified and even beyond these. Finally and perhaps most important, in terms of usable and effective deployment, is that we must ensure that we correctly and equally consider the four fundamental components of interaction and usability: user, task, system, and environment across the broadest range of possibilities for each.

3.5.2 OPPORTUNITIES AND BARRIERS TO DEPLOYMENT

Which comes first, the chicken or the egg? It is likewise with consideration of opportunities or barriers to smart technology deployment, which is it that presents first, the opportunity to do something "smart" or a barrier such as the condition or illness or situation that the client has to deal with?

Even the tendencies to see the technology as the opportunity and the person's condition as the barrier is not always correct. For many groups of people with care needs or illnesses, technology can be or can have become a barrier. As has been stated and is so often the case in the real world, opportunities and barriers are often the flip side of the same coin. Being able to toss the coin over, so to speak, is often the key to unlocking innovation and breakthrough. For example, in the case of the well-being enhancing 3D games, removal of the complex game controller (as a barrier for people with limited mobility, dexterity, and/or experience with such a device) and attempting to replace it with a device that allowed hands-free, natural, and intuitive gestures actually became a real opportunity.

Not only was the barrier of the controller resolved, but also dual player interaction proved very natural and effective, where a more able bodied and experienced "gamer" could play alongside a less able bodied, less experienced "nongamer" and together the combined interactions created more natural and intuitive game play than first thought possible!

It is not the intention here to list off in particular barriers and also opportunities to effective smart technology deployment in any comprehensive or complete sense. That would likely be a very long list! Even just considering the sorts of issues raised by UD, for example, should provide evidence of this. Technology that is truly "usable by all," "regardless of..." could generate lists of potential user classifications. "All" can and potential does mean ALL, as in everybody within a particular care setting, or care sector, or city, or region, or country or even the whole world! The user attributes, abilities (and therefore disabilities), capabilities (and therefore limitations), and capacities (and therefore incapacities) listed are extensive and can in theory represent the full array of needs, conditions, and illness possible. The real breakthrough from such barriers comes when smart technologies can be shown to work effectively for *somebody*, with a specific need in a specific situation and this then provides the opportunity for many other "somebodies" and even potentially "everybody" with such a need or condition.

Considering usability, be it as a definition, a set of design criteria or a set of evaluation metrics, in terms of barrier and/or opportunity, becomes a very useful exercise. Usability enhancement can be and often is a barrier! Overemphasis on say the learning needs of one set of users (person or group) can yield a solution that might be very frustrating, simplistic, or inflexible for another. You cannot always be all things to all users! The more sophisticated a view that is taken of what exactly constitutes usability in a given deployment context will likely provide more opportunities for design innovation. Usability is not just about "ease of use." It is also about "flexibility" and "adaptivity." Also, usability is about supporting complexity and sophistication, "hard usability" if you like. Typically people who have moderate-to-severe impairments, conditions, or illness will invest significant effort and training to become proficient or event expert at using a complex smart technology solution if that provides a valuable support.

Also we should not become too naive when trying to strive for "maximum usability" or making something "child's play" or even "idiot proof." Something even partially useful is often better than nothing, particularly in relation to care and support for "special needs" users. Some limited functionality is better than no functionality or having to wait for the provision of better functionality. Likewise, usability improvement is not by default universal. Just as very specific issues and design challenges may present in relation to different user groups or even individuals, so it will be with usability improvement. It can and often will be very clustered or even sporadic.

Another important opportunity/barrier conundrum that presents in a deployment is the availability and indeed ability of the people involved in terms of their expertise, understanding, and motivation toward dealing with usability. Obviously, it is important to have some experienced and hopefully expert input in terms of usability, human factors, and HCI more broadly. This will likely vary from project to project.

There should be a strong synergy and empathy between HCI or usability professionals and care professionals as both are centered on the user, the client, and the person. Indeed, it is most likely that from the human viewpoint, care professionals will have a lot to teach about usability to those who are coming from a more technology background!

Finally, there is the issue of being overwhelmed by the human needs that can and do present in certain care settings. This can be as a result of either the severity of the condition or need, or the scope and scale of the need. But this can be mitigated by being truly empathetic, person centered and willing to take risks on behalf of that user and that need. Necessity, even that which is severe or chronic, is always the mother of invention. Attempting to address such needs with ingenuity, practicality, and determination can be a real driver of innovation.

3.5.3 Usability and Well-Being Enhancement

As previously stated, well-being enhancement is always at least desirable and more likely the primary goal for the smart technology deployment being considered. It is always better left to care professionals (especially medical professionals in more formal health care settings) to define, measure, and attempt to improve a person's medical well-being. What is being considered here is more in terms a person's general state of well-being with a view to modeling it, to understand and deal with this complex concept from a design, development, and evaluation perspective. Typically well-being will apply to a broad set of attributes that can be classified or separated such as physical, cognitive, or emotional well-being and then within these more specific elements can be checked, such as a the number of steps walked, the correctness, or speed of recall of a piece of information or a choice selected, or even someone's generally mood. These can be added to and changed according to the particular need or set of needs being addressed and the type of care model being applied.

The key thing to remember is that "well-being" as modeled or proposed as such is very much an attribute or set of attributes of the *person* (or the user or client) and not of the smart technology being deployed! This may sound obvious but it is important to keep it in mind at all times. What is being assessed or measured or maybe even enhanced is part of a person's state of being. Usability on the other hand, particularly when we are trying to provide, or design in usability applies to and is therefore an attribute of the *technology* mainly.

Something will change in the design, construction, configuration, or deployment of the technology to enhance its usability. When we try to evaluate the effect of this change, we are then considering more of the effect on a given user in a specific situation and context. So we may for example simplify the design of a menu of options on a display and also perhaps provide different means of selecting from these. In this particular case, reducing the number or range of options may enhance the simplicity of the menu. Clustering or using color encoding or better visualization may improve the affordance offered. A touch-sensitive screen as well as arrow keys for selection may provide greater flexibility.

Visual feedback of the option selected and perhaps even confirmation of the option selected may make it more robust or less error prone, but all these objective usability enhancements have to be evaluated against realistic user profiles and tasks in a realistic setting. Ultimately it is only when it is tested by real users, in live conditions can we be sure we have a more usable menu design. The really challenging part comes when we attempt to link usability to well-being. Does usability improvement imply well-being enhancement? Also, how exactly should we "map" between usability and well-being? A more usable menu as given in the example given might be less physically demanding on a user but it might also be more time consuming to use or frustrating. Having to deal with objective measure, for example, the average number of keystroke articulations needed to find, select, and then confirm an option, alongside more subject measures such as the amount of frustration experienced are normal parts of usability testing! Every usability definition or model has to address subjective opinion, feeling, and human cost.

What is even more challenging is when we attempt to change a person's behavior to improve their well-being. This is typically the bread and butter challenge for many health care interventions. Simplifying learning, enhancing flexibility, and ensuring real robustness are all achievable from a technology deployment viewpoint, given sufficient resources and expertise, but this has to be considered alongside the much more difficult challenge of ensuring that a user remains motivated, is adequately supported, encouraged, and challenged to persist, adapt, and ultimately achieve the well-being enhancements sought.

Finally, we must never lose sight of the question of how do we know if something is more usable and that it is having the beneficial effects sought. Usability evaluation and more formal usability testing or even rigorous inspection is never really a simple undertaking. As software developers well know, if you think coding it is hard, wait until you try testing it! Add to this that designers must engage with the subjective opinions of people, and increasingly with the broader UX over the "lifetime" of engagement with the technology deployment. We are also being expected to assess the emotional responses to our designs or even to assess how intuitive a design is, hence there will never likely be any simple answers or quick fixes. But we should not despair. Usability is not an all or nothing proposition. Even with relatively poor usability, people can and will often achieve the desired goals or outcomes. It depends largely on their need in relation to the consequences of achieving success or failure. It is about making improvements, even small improvements, and measuring these against the human cost involved. As the fitness coach once quipped, "no pain no gain." It should also be remembered, however, that where there is for whatever reason(s) very poor or essentially no usability, then there will likely be no engagement, whether this is evident from either a particular individual or worse a group of individuals or indeed all users! Without reasonable and sustained engagement, a smart technology solution will have little if any lasting impact on well-being.

3.6 CONCLUSIONS

Hard problems are always hard, and quite frankly providing usable and smart technology solutions that satisfy a range of users and needs in a care setting is always going to be a hard problem to deal with effectively. When we couple this with the need to change or at least influence user behaviors, it becomes all the more difficult. In this discussion, we have quite deliberately tried to step back from the apparent "leading edge" of usability and UX technologies, tools, techniques, models, and thinking, as much of this is oriented toward technologically sophisticated, independent, capable, and discerning professionals and consumers, many of whom have grown up in the Internet age and are completely at home in the digital world of the twenty-first century. So many of the real "usability" issues that need to be dealt with within any given care setting relate to users, that is, real people, with real needs arising from real-life problems that are a normal part of the human condition and experience like aging, reduced mobility, sensory and cognitive decline, disease and sickness as well demographic barriers, lack of technological competency, literacy, or even social skills. Much of what is considered as "common sense" still applies. People are experts at being themselves, at living with such "conditions," and at learning to cope and adjust. Not only does their voice need to be heard, but also their ingenuity, resilience, problem-solving skills, and even just their grit and determination to "soldier on" needs to be embraced more empathetically.

Design empathy is important, really important! Have you ever met your "users," listened to your "users," taken time to be with your "users"? Indeed, we can ask, have you or better, exactly *how have you* helped to take care of or given care to your "users"? Usability and good care are not necessarily one and the same thing! Usability does not automatically imply well-being enhancement. Smart technologies have like any useful tool great potential. But we must carefully consider our own motivations for deploying them in a care setting. Is it more because we can, because the technology enables us to automate or monitor, or is it more because we should? Quite clearly there are important ethical and moral considerations in most all of these types of deployments, and many of these issues, although outside the scope of this discussion, are none the less important to keep in mind.

Well-being enhancement is, as was stated at the outset, one of the important goals and hopefully outcomes to be achieved in each situation. Usability has an important part to play in this but it needs careful consideration and interpretation if it is to be properly applied within care. Smart technologies offer real opportunities to enhance well-being, to improve the quality of people's lives, or to extend independence and even to mitigate or alleviate certain conditions and illnesses. But they must be deployed carefully, sensitively, and appropriately. Of course, these solutions need also to be safe, cost-effective, efficient, and dependable. But they must be learnable, even by people who find learning and relearning a challenge. They must be flexible and easily adaptable to individual needs and circumstances. They also need to be reliable, robust, and effective. They need to be up to the job in most all circumstances. We need to hold onto the core of providing real care: for people, by people to the betterment of all people.

REFERENCES

Albert, W. and Tullis, T. (2013). *Measuring the User Experience: Collecting, Analyzing, and Presenting Usability Metrics (Interactive Technologies)*. 2nd ed. Waltham, MA: Elsevier Inc.

Cairns, P. and Cox, A., eds. (2008). *Research Methods for Human-Computer Interaction*. Cambridge: Cambridge University Press.

Card, S., Moran, T., and Newell, A. (1983). *The Psychology of Human-Computer Interaction*. Hillsdale, NJ: Lawrence Erlbaum Associates Inc., pp. 23–97.

Carroll, John M. (2014). Human computer interaction—brief intro. In: Soegaard, Mads and Dam, Rikke Friis (eds.). *The Encyclopedia of Human-Computer Interaction*. 2nd ed. Aarhus, Denmark: The Interaction Design Foundation. Available online at https://www.interaction-design.org/encyclopedia/human_computer_interaction_hci.html (accessed August 10, 2015).

CEUD (Centre for Excellence in Universal Design) (2015a). *What Is Universal Design*. Available online at: http://universaldesign.ie/What-is-Universal-Design/Definition-and-Overview (accessed July 27, 2015).

CEUD (Centre for Excellence in Universal Design) (2015b). *The 7 Principles*. Available online at http://universaldesign.ie/What-is-Universal-Design/The-7-Principles (accessed July 27, 2015).

CUD (Center for Universal Design) (2008). *About UD*. Available online at: http://www.ncsu.edu/ncsu/design/cud/about_ud/about_ud.htm (accessed August 11, 2015).

Dix, A., Finlay, J., Abowd, G. D., and Beale, R. (2004). *Human-Computer Interaction*. 3rd ed. Harlow: Pearson Education, pp. 260–273.

Finn, E., Sirkka, A., Merilampi, S., Leino, M., and Koivisto, A. (2014). Exploring interactive gameplay for well-being enhancement. *Journal of Finnish Universities of Applied Sciences*, 3. Available online at: https://arkisto.uasjournal.fi/uasjournal_2014-3/finn_merilampi.html (accessed August 15, 2016).

Finn, E., Sirkka, A., Merilampi, S., Leino, M., and Koivisto, A. (2015). Cognitive mobile games for memory impaired adults: An Irish comparative case study. In: *Satakunta University of Applied Sciences Series B, Reports 3/2015*. pp. 21–26. Pori, Finland: SAMK.

Gould, J. D. and Lewis, C. (1985). Designing for usability: Key principles and what designers think. *Communications of the ACM*, 28(3): 300–311.

IGBL (Irish Symposium on Game Based Learning) (2015). Haven: A configurable simulation game designed to elicit an emotional response. Available online at: https://igblconference.wordpress.com/presenters/ (accessed August 15, 2015).

ISO (1998). *ISO 9241-11:1998(en) Part 11: Guidance on Usability*. Available online at: https://www.iso.org/obp/ui/#iso:std:iso:9241:-11:ed-1:v1:en (accessed August 11, 2015).

Karwowski, W., Soares, M., and Stanton, N., eds. (2011). *Human Factors and Ergonomics in Consumer Product Design, Methods and Techniques.* Boca Raton, FL: CRC Press, Taylor & Francis Group.

Kiili, K., Ketamo, H., Koivisto, A., and Finn, E. (2014). Studying the user experience of a tablet based math game. *International Journal of Game-Based Learning*, 4(1), 60–77, January–March.

Koivisto, A., Sirkka, A., and Merilampi, S. (2015). Activation and rehabilitation games for people with special needs. In: *Satakunta University of Applied Sciences Series B, Reports 3/2015.* pp. 27–32. Pori, Finland: SAMK.

Leap (2015). *Leap Motion Controller: Hand Sensor for VR and AR Applications.* Available online at: https://www.leapmotion.com (accessed August 24, 2015).

McGonigal, J. (2011). *Reality Is Broken. Why Games Make Us Better and How They Can Change the World* [Kindle]. London: The Random House Group Ltd. Epub.

Mega (2015). *Mega Future Games.* Available online at: http://www.megafuturegames.net (accessed August 24, 2015).

Merilampi, S., Sirkka, A., Leino, M., Koivisto, A., and Finn, E. (2014). Cognitive mobile games for memory impaired older adults. *Journal of Assistive Technologies*, 8(4), 207–223.

Nielsen, J. (1993). *Usability Engineering.* London: Academic Press.

Nielson, J. (2012). *Norman Nielsen Group—Usability 101: Introduction to Usability.* Available online at: http://www.nngroup.com/articles/usability-101-introduction-to-usability (accessed August 8, 2015).

Nielsen, J. and Budiu, R. (2013). *Mobile Usability.* Berkeley, CA: New Risers, pp. 9–44.

NNG (Norman Nielsen Group) (2015). *Evidence-Based User Experience Research, Training, and Consulting.* Available online at: http://www.nngroup.com (accessed August 20, 2015).

Oculus (2015). *Oculus Rift: Next-generation Virtual Reality.* Available online at: https://www.oculus.com/en-us/rift (accessed August 24, 2015).

PIP (2015). *The Pip Stress Management Device Measures Electro Dermal Activity (EDA).* Available online at: https://thepip.com/en-eu (accessed on August 25, 2015).

Rubin, J. and Chisnell, D. (2008). *Handbook of Usability Testing: How to Plan, Design, and Conduct Effective Tests* [Kindle]. 2nd ed. Indianapolis, IN: Wiley Publishing, Inc., Chapter 1.

Ryan, P. and Finn, E. (2005). Field-based mLearning: Who wants what? In: Isaias, P., Borg, C., Kommers, P., Bonanno, P. (eds.). *Proceedings of the IADIS International Conference— Mobile Learning 2005*, Qwara, Malta. pp. 256–260.

Scrumwise (2015). *Scrumwise: Better Scrum.* Available online at: https://www.scrumwise.com/features.html (accessed August 12, 2015).

Shackel, B. (1986). Ergonomics in design for usability. In: Harrison, M.D and Monk, A.F. (eds.), People and computers: Designing for usability, *Proceedings of the Second Conference of the British Computer Society, Human Computer Interaction Specialist Group.* pp. 44–64, Cambridge, UK: Cambridge University Press.

SIGCHI (1992). *ACM SIGCHI Curricula for Human-Computer Interaction.* Available online at: http://old.sigchi.org/cdg/cdg2.html#2_1 (accessed August 10, 2015).

Sirkka, A. and Koivisto, A. (2015). Art and technology play well together. In: Sirkka, A (ed.), *Satakunta University of Applied Sciences Series B, Reports 3/2015.* pp. 10–16. Pori, Finland: SAMK.

Usability (2015). *Improving the User Experience.* Available online at: http://www.usability.gov (accessed August 22, 2015).

4 IP Strategies in eHealth and eCare Sectors

Díez-Díaz Mónica, López-Moya J. Rafael,
and Díez-Díaz Alvaro

CONTENTS

4.1 INTRODUCTION

One of the major objectives of the technology companies is to be the first in the market with new products/processes that meet the market needs, to achieve a competitive advantage meaning, the highest market penetration by enhancing the target size or by excluding competitors. Other advantages apart from gaining access to technology are the opportunities to build joint ventures, to create stable relationships with suppliers and customers, to produce revenue, directly and indirectly, and to establish and protect brand value.

The main challenge to the success of technological companies/institutions depends on their ability to manage intangible assets or to transfer intellectual property rights (IPRs) at the appropriate value. Managing and licensing technological innovations is a complex and multidisciplinary task. Various parties with different expertise need to deal with one another at multiple, complex levels, simultaneously and for mutual benefit. The technology market is segmented into those who create tools and technologies and those who develop and commercialize products that use those tools and technologies [1].

To ensure that those who create the tools and technologies keep their IPR once the products are launched, there are different forms of protection that help us to achieve the objective: Patents, Copyright and Related Rights, Trademarks, Geographical Indications, Industrial Designs, Layout Designs (Topographies) of Integrated Circuits (ICs), Protection of Undisclosed Information, Utility Models, Design Patents, Industrial Designs, Trade Secrets, Domain Names, and Licensing agreements.

Patents play an important role enabling companies to develop, for example, life-related products, due to the fact that the patent confers the IPR for a limited period of time, providing the opportunity to recover the investment made in developing the patented invention, launching the new products into the market, or licensing or trans-ferring said rights to other companies. Therefore, the exploitation of IPR, patents, software, etc. is necessary to create more opportunities in the society, providing an incentive for innovation. On the other hand, inadequate intellectual property protec-tion and management is one of the major barriers to commercialization especially in, eHealth, eCare, eMonitoring AAL technologies, and well-being sector.

Several authors propose that a powerful indicator that entrepreneurial firms have articulated the project's potential usefulness and commercial viability [2] is applying for patenting the product idea/project as it can accelerate collaboration. This entails a reduction in uncertainty, which is important in the early stages of a project, such as, for example, in the start-up creation. According to Katila and Mang [2], those firms that have applied for a patent in a project are likely to have a high R&D intensity, mutual collaboration experience with a partner, or prior general R&D collaboration experience.

4.2   IPRs IN THE eHEALTH AND eCARE FIELDS

To present the IPRs in the subject of interest eHealth and eCare fields, particularly focused on Active Assisted Living Solutions and Well-being, we need to introduce wider concepts on the IP trade-related subject.

The Trade-Related Aspects of Intellectual Property Rights (TRIPS) [3] is an international agreement governed by the World Trade Organization (WTO) [4], which sets down minimum standards for many forms of IP regulation as applied to nationals of WTO members.

The term *intellectual property rights* (IPRs) refers to all categories of intellectual property such as (1) Copyright and Related Rights, (2) Trademarks, (3) Geographical Indications, (4) Industrial Designs, (5) Patents, (6) Layout Designs (Topographies) of Integrated Circuits, and (7) Protection of Undisclosed Information.

Computer programs, whether in source or object code, shall be protected under the Berne Convention (1971) [5], and compilations of data or other material, whether in machine readable or other form, which by reason of the selection or arrangement of their contents constitute intellectual creations shall be protected as such, without prejudice to any copyright subsisting in the data or material itself. But members shall confine limitations or exceptions to exclusive rights to certain special cases that do not conflict with a normal exploitation of the work and do not unreasonably prejudice the legitimate interests of the right holder. And this is the case of the US Patent Act [6].

According to TRIPS, patents shall be available for all inventions, whether they are products or processes, in all fields of technology, provided that they are new, involve an inventive step, and are capable of industrial application, without discrimination as to the place of invention, the field of technology, and whether the products are imported or locally produced. All these rights are enjoyable if the subject matter is considered patent eligible, according to the US Patent Act, or if it is considered as an invention, according to the European Patent Convention (EPC) [7].

Essentially, the US patent system limits exclusions from patent eligibility to three items: (i) laws of nature, (ii) natural phenomena, and (iii) abstract ideas [8]. Traditionally, the USA (and also Canada, Korea, and Japan) was the most tolerant country with the protection of inventions related with software by means of a patent, but the *Bilski V. Kappos* sentence in 2010 [9] supposed an important limitation in respect to the patent eligibility in the USA recognizing that a computer program is a patent eligible if the process associated with the program is directly related to a machine or determines the transformation of an item into another thing or state.

According to the EPC, European patents shall be granted for any inventions, with explicit exception of (a) discoveries, scientific theories, and mathematical methods; (b) aesthetic creations; (c) schemes, rules, and methods for performing mental acts, playing games or doing business, and programs for computers; and (d) presentations of information. Software is not considered as an invention but may be protected in the form of a "computer implemented invention"; if it is running in a computer, this causes an additional technical effect that goes beyond the normal physical interaction between software and hardware.

To protect the invention by means of a patent, the next step is to determine if the invention is excluded or not from patentability. According to TRIPS, inventions are excluded if they (i) prevent commercial exploitation to protect public order or morality, including to protect human, animal, or plant life or health or to avoid serious prejudice to the environment, provided that such an exclusion is not made merely because the exploitation is prohibited by their law, or (ii) to protect the competence of physicians and classical plant breeders, for example, when the invention is a diagnostic, therapeutic, or surgical method for the treatment of humans or animals, or plants and animals other than microorganisms, and, essentially, biological processes for the production of plants or animals other than nonbiological and microbiological processes. Also in this matter, the US Patent Act and the EPC have different limitations. The US Patent Act accepts inventions referred to diagnostic, surgical, or treatment methods that developed directly on the human or animal body. In contrast, in EPC, the invention must be developed outside the human or animal body, or in particular diagnostic methods, if intervention on body is needed, with the proviso that the life of the patient is not at risk. This exclusion in the EPC is based on socio-ethical and public health considerations, with the intention to ensure that nobody who wants to use the methods as part of the medical treatment of humans or animals should be prevented from this by patents. In other words, the exclusion intends to protect patients. This exception of patentability according to the EPC shall not apply to products for use in diagnostic, surgical, or treatment methods.

Because patentability issues are very controversial in areas related with biomedicine, there is a great continuous effort on the part of patent offices, for example, the European Patent Office (EPO), to implement this legislation in their daily work and adapt the content of their legal framework to the actual situation of the society. The Court of Justice of the EU recently gave its opinion on the limits to patentability for human embryonic stem cells. The EPO will immediately implement a restrictive interpretation of biotech patentability provisions according to the court's decisions. The proof is that the 1998 biotechnology directive was incorporated into the EPO's legal framework in 1999, much earlier than in many EU member states.

AAL Technologies, Smart eHealth, and eCare Technologies are an emerging sector closely supported by electronic processes and communication, covering electronic/digital processes in health and care. The IPR of this sector is related to the next modalities.

4.2.1 COPYRIGHT

Computer programs are protected as literary works within the meaning of Article 2 of the Berne Convention [5]. Such protection applies to computer programs, whatever may be the mode or form of their expression. Then, as indicated previously, the software, specifically the "lines of code" could be protected by copyright. Also, the technical and preparatory documentation and the user's manuals can be protected by copyright, in the adequate category.

The structure and content of the databases (e.g., electronic health records) of the software would also be registered as copyright.

All the software, structure, and content of the databases could be used to several objectives:

- *Health care information systems:* Software solutions for appointment scheduling, patient data management, work schedule management, and other administrative tasks surrounding health.
- *Computerized physician order entry:* A means of requesting diagnostic tests and treatments electronically and receiving the results.
- *ePrescribing:* Access to prescribing options, printing prescriptions to patients and sometimes electronic transmission of prescriptions from doctors to pharmacists.
- *Clinical decision support:* Providing information electronically about protocols and standards for health care professionals to use in diagnosing and treating patients.
- *Virtual health care teams:* Consisting of health care professionals who collaborate and share information on patients through digital equipment.
- *mHealth:* Comprises the use of mobile devices in collecting aggregate and patient level health data, providing health care information to practitioners, researchers, and patients, real-time monitoring of patient vitals, and direct provision of care.
- *Health knowledge management:* In an overview of latest medical journals, best practice guidelines, or epidemiological tracking.

- *Medical research using grids:* Powerful computing and data management capabilities to handle large amounts of heterogeneous data.
- *Electronic health records:* Enabling the communication of patient data between different health care professionals.

The term of protection granted by this Berne Convention shall be the life of the author and 70 years after his death, in all contracting states (nowadays 168 states).

4.2.2 PATENTS

Taking into account the patentability requirements of the corresponding state, the process involving software and hardware with an additional technical effect can be protected by means of a patent, for example, in the United States Patent and Trademark Office (USPTO) or in the case of EPO under computer implemented inventions.

Specifically, the particular devices used to be applied on the eHealth and eCare can be protected by a patent. For example, telemedicine can be a clear application of these devices:

- *Telemedicine:* physical and psychological diagnosis and treatments at a distance, including telemonitoring of patient functions
- Mobile telemedicine

Usually the patent system provides 20 years of protection for the invention, as in the USA and Europe. The protection has a nature of territoriality because every state examines the patentability of the invention and grants it for the particular state or, in the case of regional agreement, for all the member states of the agreement, as in the case of the European Patent Office (EPO), African Regional Intellectual Property Organization, African Intellectual Property Organization, Eurasian Patent Organization, and Gulf Cooperation Council.

4.2.3 INDUSTRIAL DESIGNS

According to Directive 98/71/EC of the European Parliament and the Council of October 13, 1998 [10], on the legal protection of designs, a design is defined as "the appearance of the whole or a part of a product resulting from the features of, in particular, the lines, contours, colors, shape, texture and/or materials of the product itself and/or its ornamentation."

Registered and unregistered community designs are available under EU Regulation 6/2002 [11], which provide a unitary right covering the European Community. Protection for a registered community design is initially valid for 5 years from the date of filling up to 25 years, subject to the payment of renewal fees every five years. The unregistered community design lasts for three years after a design is made available to the public and infringement only occurs if the protected design has been copied exactly.

According to the US system, the design can be registered as a "design patent." The design consists of the visual ornamental characteristics embodied in, or applied to, an article of manufacture. Because a design is manifested in appearance, the subject matter of a design patent application may relate to the configuration or shape of an article, to the surface ornamentation applied to an article, or to the combination of configuration and surface ornamentation. A design for surface ornamentation is inseparable from the article to which it is applied and cannot exist alone. It must be a definite pattern of surface ornamentation, applied to an article of manufacture. A design patent protects only the appearance of the article and not structural or utilitarian features. Patents for designs shall be granted for the term of 14 years from the date of grant.

4.2.4 TRADEMARKS

The association between a denomination, graphic, or sound and any product or service, or even enterprise denomination (in the case of Spain, "Trade Name") could be protected as the trademark system that any state has been adopted. Trademarks are signs (capable of being represented graphically) used in trade to identify products and services. According to Directive 2008/95/EC of the European Parliament and of the Council of October 22, 2008 [12], to approximate the laws of the member states relating to trade marks, a trade mark may consist of any signs capable of being represented graphically, particularly words, including personal names, designs, letters, numerals, the shape of goods or of their packaging, provided that such signs are capable of distinguishing the goods or services of one undertaking from those of other undertakings.

Each registration shall remain in force for 10 years. Trademarks can be renewed every 10 years indefinitely.

4.2.5 LAYOUT DESIGNS (TOPOGRAPHIES) OF INTEGRATED CIRCUITS

This category of IPR protection is a *sui generis* modality of protection. The objective of this category is the protection of ICs, chip, or microchip. The IC is the combination of electronic elements in a particular layout but is not related to a technical effect; it is not considered as an invention and cannot be protected as a patent (or utility model) nor as industrial design because of the lack of aesthetic means.

A diplomatic conference was held at Washington, DC, in 1989, which adopted a Treaty on Intellectual Property in Respect of Integrated Circuits [13], also called the Washington Treaty or IPIC Treaty. The IPIC Treaty is currently not in force, but has been incorporated by reference into the TRIPS agreement.

"Integrated circuit" means a product, in its final form or an intermediate form, in which the elements, at least one of which is an active element, and some or all of the interconnections are integrally formed in and/or on a piece of material and which is intended to perform an electronic function.

"Layout design (topography)" means the three-dimensional disposition, however expressed, of the elements, at least one of which is an active element, and some or all of the interconnections of an IC, or such a three-dimensional disposition prepared for an IC intended for manufacture.

The exclusive right of the owner also extends to articles incorporating ICs in which a protected layout design is incorporated, insofar as it continues to contain an unlawfully reproduced layout design; the circumstances in which layout designs may be used without the consent of owners are more restricted.

4.2.6 PROTECTION OF UNDISCLOSED INFORMATION

A trade secret is valuable information for an enterprise that is treated as confidential and that gives the enterprise a competitive advantage.

The European Commission is working to harmonize the existing diverging national laws on the protection against the misappropriation of trade secrets so that companies can exploit and share their trade secrets with privileged business partners across the internal market, turning their innovative ideas into growth and jobs.

In November 2013, the Commission proposed a draft directive [14] that will align existing laws against the misappropriation of trade secrets across the EU. The proposal is now being examined by the European Parliament.

The draft directive does not provide any grounds for companies to hide information that they are obliged to submit to regulatory authorities or to the public at large. Moreover, the draft directive does not alter and does not have any impact on the regulations that foresee the right of citizens to access documents in the possession of public authorities, including documents submitted by third parties such as companies and business organizations.

4.3 CURRENT SITUATION IN THE PROTECTION BY PATENT IN THE eHEALTH OR eCARE FIELDS

To evaluate the evolution and the current scenario of AAL Technologies, eHealth, eCare, and well-being fields related to information and communications technology (ITC) using the protection of the inventions by patent as the reference indicator, a search has been carried out using the terms [telecare or telehealth or telemonitoring or eHealth or eCare or ageing or "elderly people" or gerontology or longevity or (management and "chronic conditions") or ("social inclusion") or (access and "self-serve society") or (mobility and "older adults") or (management and "daily activities") or ITC or "Information and communications technology" or (well-being)] in title, abstract, and claims for the period of time from 1996 to 2015 of the Derwent World Patents Index (DWPI) indexed patent or patent applications. DWPI is the world's most comprehensive database of enhanced patent documents.

Because of the number of terms studied and their general and wide impact in patents of all different natures, preliminary searches were performed to refine the results. One of the tools to reduce the spectrum was the mapping of the patents per families. The analysis demonstrates that some IPCs (International Patent Classification) such as antiaging cosmetics or classical pharma products were scattered in the results and misleading the analysis from our interest, that is, eHealth, eCare, AAL Technologies, and self-serve society.

Based on the findings exposed, further searches were performed with the relevant fields and excluding the references to traditional pharma and cosmetics; 2413 family patents were localized and analyzed; the patent publishing trend from 1996, top countries, top IPC, and assignees (top assignees, top assignees by year, and assignee distribution). The relevance of the study is ensured by covering the last 20 years of data.

In Figures 4.1, 4.3 through 4.5, the Y-axis shows document count. Figure 4.1 shows the progression of the number of patent applications published along all the represented period. The steady steep increase on the protection of AAL Technologies, eHealth, eCare, and self-serve society-related technologies is notable; there is an exponential increase from 2005 to 2011 that becomes steeper in the following years. This is an indicator of the expansion and commercial interest of said technologies.

The USA and China stand out by the number of patent applications filed in these countries, representing 55% of all the patent applications. The third place is for the WO: World Intellectual Property Organization followed by the EP: European Patent Office and Japan (JP) (Figure 4.2) (US: United States of America, CN: China, WO: World Intellectual Property Organization, EP: European Patent Office, JP: Japan, AU: Australia, KR: Republic of Korea, CA: Canada, GB: Great Britain, DE: Germany, RO: Romania, FR: France, IT: Italy, TW: Taiwan, IN: India, ES: Spain, RU: Russia, BR: Brazil, NL: the Netherlands, SG: Singapore).

The IPC, established by the Strasbourg Agreement 1971 [15], provides a hierarchical system of language-independent symbols for the classification of patents and utility models according to the different areas of technology to which they pertain. Most of the patents filed in the AAL Technologies, eHealth, eCare, well-being, and self-serve society fields refer to:

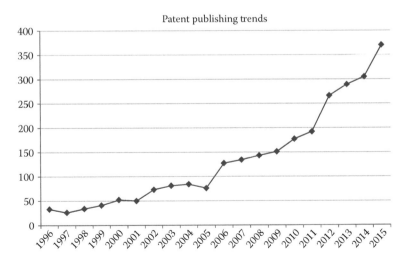

FIGURE 4.1 Number of patent applications related to AAL Technologies, eHealth, eCare, and self-serve society-related technologies published along all the represented period (1996–2015).

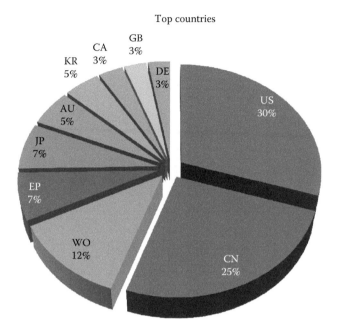

FIGURE 4.2 Number of patent applications per country related to AAL Technologies, eHealth, eCare, and self-serve society-related technologies published along all the represented period (1996–2015).

- Data processing systems or methods, specially adapted for administrative, commercial, financial, managerial, supervisory, or forecasting purposes
- Arrangements, apparatus, circuits, or systems characterized by a protocol
- Alarms responsive to unspecified undesired or abnormal conditions measuring for surgery, diagnostic purposes, and identification of persons
- Chemical investigation or analysis of biological material
- Educational or demonstration appliances for teaching or communicating to mitigate human needs and limitations

These are the subject matter of interest to be protected by patent (Figure 4.3 and Table 4.1).

As is shown in Figure 4.4 and Table 4.2, the top 15 in the patent activity in the sector are represented (in order of number of published patents) by

Li Zong Cheng; Accenture LLP; Matsushita Electric Ind Co. Ltd; Nestec SA; Boeing Co.; Koninkl Philips NV; Ericsson Telefon AB LM; Oreal; Raniero Seri; Schlumberger Technology Corp.; Casio Computer Co. Ltd; Searete LLC; IBM; Univ. California; Tunstall Group Ltd.

It is remarkable that the presence of Mr. Zong Cheng Li, who is a professor from Su Zhou University (China), tops the list with 67 patent families in the top 15 assignees.

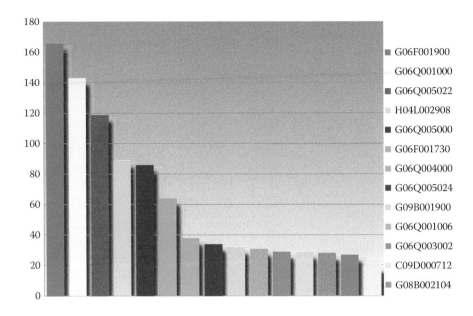

FIGURE 4.3   Top 15 IPCs per number of patent applications related to AAL Technologies, eHealth, eCare, and self-serve society-related technologies published along all the represented period (1996–2015).

TABLE 4.1
Distribution of the International Patent Classification Related to eHealth, eCare, Well-Being and Self-Serve Society Fields

Current IPC	Description	Document Count	Percentage
G06F001900	Electric digital data processing. Architectures	166	16
G06Q001000	Data processing systems or methods, specially adapted for administrative, commercial, financial, managerial, supervisory, or forecasting purposes. The special adaptation is determined to be novel and nonobvious	143	14
G06Q005022	Data processing systems or methods, specially adapted for administrative, commercial, financial, managerial, supervisory, or forecasting purposes. Health care	119	12
H04L002908	Arrangements, apparatus, circuits, or systems for transmission control procedure	89	9
G06Q005000	Data processing systems or methods, specially adapted for administrative, commercial, financial, managerial, supervisory, or forecasting purposes. Utilities	86	8

(*Continued*)

TABLE 4.1 (*Continued*)

Distribution of the International Patent Classification Related to eHealth, eCare, Well-Being and Self-Serve Society Fields

Current IPC	Description	Document Count	Percentage
G06F001730	Electric digital data processing. Information retrieval	64	6
G06Q004000	Data processing systems or methods, specially adapted for administrative, commercial, financial, managerial, supervisory, or forecasting purposes. Systems or methods that involve significant data processing operations. Finance; insurance	38	4
G06Q005024	Data processing systems or methods, specially adapted for administrative, commercial, financial, managerial, supervisory, or forecasting purposes. Systems or methods that involve significant data processing operations. Systems or methods that are specially adapted for a specific business	34	3
G09B001900	Educational or demonstration appliances; appliances for teaching or communicating with the blind, deaf, or mute; models; planetaria; globes; maps; and diagrams. Teaching or practice apparatus	32	3
G06Q001006	Data processing systems or methods, market research and analysis, surveying	31	3
G06Q003002	Data processing systems or methods, resources, workflows, human, or project management	29	3
C09D000712	Features of coating compositions	29	3
G08B002104	Alarms responsive to a single specified undesired or abnormal condition. Responsive to nonactivity	28	3
G06Q003000	Data processing systems or methods, resources, workflows, human, or project management. Commerce	27	3
H04L002906	Arrangements, apparatus, circuits, or systems characterized by a protocol	20	2
G08B002300	Alarms responsive to unspecified undesired or abnormal conditions	19	2
G06Q001010	Data processing systems or methods, specially adapted for administrative, commercial, financial, managerial, supervisory, or forecasting purposes. Systems or methods that involve significant data processing operations. Office automation	18	2
G08B002102	Alarms responsive for ensuring the safety of persons	18	2
G01N003350	Chemical investigation or analysis of biological material	17	2
G01N003368	Investigating or analyzing materials by specific methods	16	2

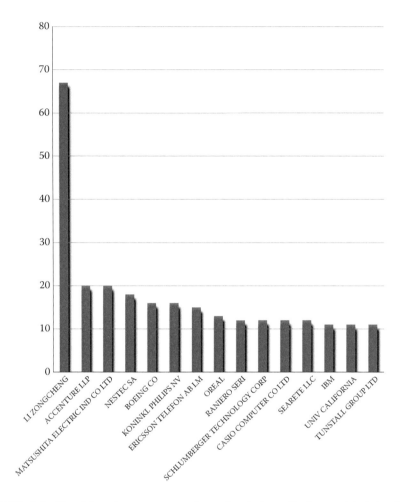

FIGURE 4.4 Top 15 assignees per number of patent applications related to AAL Technologies, eHealth, eCare, and self-serve society-related technologies published along all the represented period (1996–2015).

It is even more remarkable that Mr. Zong Cheng Li is only applying for local Chinese patents without the option to extend the patent rights to additional countries; on the contrary, the rest of the contributors are opting for a wider scope of countries and more aggressively building comprehensive patent families. Top five players (excluding Mr. Zong Cheng Li) have extended their patents to the United States, World Intellectual Property Organization, European Patent Office (EP), and most to Australia, Canada, and China.

Analyzing the distribution of the published patents between the top 15 assignees with the exception of Mr. Zong Cheng Li, there is noticeable proportionality in the distribution (Figure 4.5). Based on this, proportionality can be understood that the

TABLE 4.2
Number of Published Patents (or Patent Applications) Filed by the Top 15 World Assignees

Assignee/Applicant	Document Count	Percentage
Li Zong Cheng	67	25.19
Accenture LLP	20	7.52
Matsushita Electric Ind. Co. Ltd	20	7.52
Nestec SA	18	6.77
Boeing Co.	16	6.02
Koninkl Philips NV	16	6.02
Ericsson Telefon AB LM	15	5.64
Oreal	13	4.89
Raniero Seri	12	4.51
Schlumberger Technology Corp.	12	4.51
Casio Computer Co. Ltd	12	4.51
Searete LLC	12	4.51
IBM	11	4.14
Univ. California	11	4.14
Tunstall Group Ltd	11	4.14

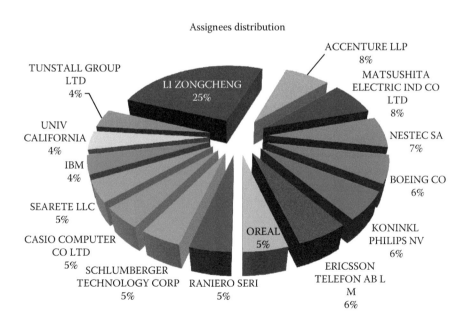

FIGURE 4.5 Top 15 assignees per percentage of patent applications related to AAL Technologies, eHealth, eCare, and self-serve society-related technologies published along all the represented period (1996–2015).

sector is opening fair opportunities to new competitors with a good patent strategy to compete with current players.

4.4 KEY POINTS IN THE IPR STRATEGY

Analyzing the results obtained in the previous point, it is appropriate to introduce the term IPR strategy as a tool every company can develop and as a key factor for competence and survival specially referring to the terms of interest in this chapter, AAL Technologies, eHealth, eCare, and self-serve society, where new technologies play a significant role, but not only because in the current interconnected world a solid IPR strategy is more than advisable for any company in any sector that aims to be competitive and successful. For that reason, the concepts developed below are not just applicable to the scope of the study but to all companies who strive for success.

In real life, many companies are focused on creating and obtaining patents but managing effectively the portfolio is even more crucial. Managing patents strategically will have an impact on bottom to top line of revenue. By streamlining the patent portfolio on one hand, savings will increase avoiding unnecessary cost and on the other hand, incomes will also increase by aligning with the company strategy.

There is a clear positive correlation between strong IP rights and innovation. Innovation is most active in countries where IP systems have been well established for a long period of time.

There are three easy steps to create value from intellectual assets [16]: (1) finding the subject matter of interest in a technical field, the intellectual asset business problem; (2) making sure that the solution to the problem has an economic impact shows its value in the market and excludes competitors; and (3) repeating steps 1 and 2.

It is necessary to think globally, incorporating a global perspective with a clear understanding of local conditions. Companies develop IP to obtain several advantages [17]:

1. To gain access to technology, for example, be the first to protect a technology or through cross-licensing
2. To build joint ventures
3. To create stable relationships with suppliers and customers
4. To produce revenue, directly and indirectly
5. To help to establish and protect brand value

Intellectual capital (IC) is composed of (1) human capital, (2) structural capital (IC, patents included), and (3) relational capital [18]. The management of IC is essential to ensure the survival of any company in the middle/long term. A very good explanation of the development of IC management capacity is the IC pyramid [19]: (1) a defense strategy at the base of the pyramid, (2) cost control, (3) revenue generation, (4) integration of (1) to (3), and (5) visionary management of IC, at the top of the pyramid. A good statement about visionary management would be "the best way to predict the future is to invent it" (George Pake, founder of Xerox PARC). A visionary company can create IC that will influence the market for years. Old manufacturing companies, for example, Dutch State Mines (DSM), can also act as IP visionaries. DSM reinvented itself to become a life and material sciences company.

IC pyramid [19]:

(5) Visionary management of IC, at the top of the pyramid. Positioned for technological revolution and growing the company future market share, IP as functional excellence, IP priorities broadened and aligned with company-wide pursuits with value extraction.

(4) Integration of (1) to (3). IP present at daily activities, focus on process not just assets, IP has focus across technology and legal.

(3) Growth-driven revenue generation, assets aligned to growth plans, possible monetization, and licensing, IP considered as business asset.

(2) Cost control, cost conscious maintaining core assets, IP considered as legal asset.

(1) Defense strategy at the base of the pyramid. Stakes claims and builds protections, IP considered as legal asset.

IP's Strategic Impact and Involvement:

Establishing the four strategies to follow, how to interact between them, and which tools will support decisions and measurements (Table 4.3).

The common situation of IP management in an institution/company is as follows: (1) occasionally, the emergence of an idea; (2) investigating, developing, and protecting the invention; (3) setting up and managing the portfolio of intangible assets; and (4) finding utility for some of the intangible assets.

TABLE 4.3
IP's Strategic Impact and Involvement

Strategy II	Strategy IV
Limit competitors	Leverage full potential
Strategy I	Strategy III
Protect intangibles	Establish income stream

Tools:
State of the art
Freedom to operate
Right of use
Design around
Clearance
Competitive IP assessment
Patentability and asset development
Problem solving—internal, customer, and so on
Portfolio development
Standards and pooling
Assertion practice
IP appetite versus risk: low, medium, high

The preferred IP management would be the next scenario:

1. Finding opportunities in the technological market and creating the idea by exploring several methods,
2. Investigating, developing, and protecting the invention,
3. Setting up and managing the portfolio of intangible assets, and
4. Negotiating the value of all of the IP.

In the common IP management, the team manager analyzes the market and identifies the opportunities in a technology field as the main task at the end of the IP management, whereas in the preferred IP management, these tasks move to the first stage of the management process, when the idea is still not defined by the promoters. In this way, the human team joins forces and, as a result, the institution/company optimizes resources.

All the patentable inventions enrolled in the common IP management are part of a wide portfolio after many years have elapsed, which entails high and unsustainable costs. On the contrary, with the preferred IP management, the portfolio is set up by inventions created to fill a particular gap in the market meaning that assets will be aligned with growth plans. Therefore, the costs generated will be reduced to the core assets and in the optimal scenario will be more than sustainable. An additional advantage of the optimal scenario is that significant revenue is reimbursed to the source, by identifying more opportunities in the market, which means feedback to the IP management. Another advantage of having strong IP in the portfolio is that can serve as a nucleus to create a start-up, attract venture capital, or cross-license. Although occasionally a "weak" patent, or patent application, has been licensed or received a huge amount of revenue to its proprietors, the desirable situation of any proprietor is to have a strong IP portfolio.

In Table 4.4, we present key factors to integrate the preferred IP management with the growth strategy, covering wider concepts within the company.

TABLE 4.4
Integrated IP Growth Strategy

Business strategy	Facts to consider:
Technology/marketing strategy	Alignment with business strategy
Tax strategy	IP mindset
IP strategy	Competitively asses IP Opportunities/risk
	How to leverage protection timing
	Defining core IP processes, resources and priorities
	Reduce cycle time and improved efficiencies
	Constant value creation and opportunity leverage

4.5   OTHER DETERMINANTS IN THE IPR STRATEGY

To commercialize a process or product, there are some aspects that must be considered.

4.5.1   MARKET NEED

There is a clear positive correlation between strong IP rights and innovation, and both are involved in a feedback contribution.

Any potential market need represents an opportunity for any company, the challenge is to find it. This opportunity can allow an operation without competitors. To achieve success in this task, there are competitive intelligent tools that inform the company about the presence and size of gaps to be filled or specific technologies with a limited number of competitors. The complement to this strategy is to predict the behavior of the located need within the context of the technology area of interest, for example, by means of competitive intelligence or technological surveillance. In this regard, the best way to predict the future of any technology is the visionary mode. In fact, visionary management makes it possible to create intellectual capital that will influence the market for years, but, according to the IC pyramid [19], this usually occurs after the defense strategy, cost control, revenue generation, and the integration thereof. Therefore, an acceptable option to profit from the opportunities of market need is the scenario with the preferred IP management described above.

4.5.2   FREEDOM TO OPERATE ANALYSIS

Patent rights allow the owner to prevent third parties from appropriating the subject matter covered by the claims of the granted patent. However, it does not give the owner any specific right to exploit the patent himself/herself (offer, use, introduce, or manufacture it). This is the called the "negative rights" conferred by the patent.

Before the owner can exploit the subject matter covered by the patent, she/he must take into consideration (i) any patent rights of third parties and (ii) any other laws applying to the activity she/he is proposing, for instance: legislation on the commercialization of monitoring health care devices on a specific country or region, legislation concerning the eHealth data exchange, the conditions of access to the different free software (four freedoms), and so on.

Therefore, freedom to operate (FTO) analysis is a necessary tool to delimit the scope of marketing of a product/process or the exploitation of any patent rights. Because of the different restrictions and IP rights in different jurisdictions, an FTO analysis should focus on particular countries or regions to operate.

4.5.3   EFFICIENCY IN PROSECUTION BY THE PATENT OFFICES

Efficiency in prosecution supports commercial innovation and addresses global challenges in the market. The main improvements recently achieved are, for example, a patent reform at the USPTO, accelerated examination of green technology applications at the EPO and USPTO, and initiatives by the World Intellectual Property

Organization using the Patent Cooperation Treaty as a platform for work sharing among major patent offices.

Another very useful tool for owners is the program for accelerated prosecution of European patent applications (PACE program) [16], which enables applicants to obtain the search report, the first examination report and any communication under Rule 71(3) EPC within tight deadlines [20].

A crucial factor in this regard is the legal framework with respect to IPR in developing countries such as China or India. The harmonization of their property rights systems with those of developed countries will be a scenario that will encourage the development of these countries and ensure the growth of the economy around the world.

4.6   CONCLUSIONS

Needless to say that new technologies such as data processing, data analysis, and cloud services are revolutionizing AAL, eHealth, or eCare, and well-being fields related to ITC that will have deep impact boosting the quality of human life. These new technologies are opening the sector to new key players by tradition not focused on health but rather in consultancy services or mobile technologies fostering the possibilities into new lines of business. In any case to fully tap the market potential, players will need to implement an effective IPR strategy taking into account the appropriate modalities of IPR protection for each achieved result.

When patents are selected as part of the IPR protection strategy, it is advisable to have a strong patent portfolio both to protect oneself and to attack any patent infringers with the strongest possible guarantees of success. As mentioned in Section 4.4, managing effectively the portfolio is even more crucial because it will have an impact on bottom to top line of revenue. Additional factors such as FTO or the efficiency in prosecution by the patent offices are essential to have an effective management of the intangible assets of the company.

REFERENCES

1. Leslie GR. Perspectives on successful biotechnology licensing Part 1: The basics. *BioProcess International* 2005, 3(8): 28–30.
2. Mang PY, Katila R. Exploiting technological opportunities: The timing of collaborations. *Research Policy* 2003, 32: 317–332.
3. World Trade Organization (WTO). TRIPS. [Online] January 1, 1995. https://www.wto.org/english/tratop_e/trips_e/trips_e.htm (accessed on August 19, 2016).
4. World Trade Organization (WTO) World Trade Organisation Official Webpage [Online] January 1, 1995. https://www.wto.org/ (accessed on August 19, 2016).
5. World Intellectual Property Organization (WIPO). Berne Convention. [Online] 1971. http://www.wipo.int/treaties/en/text.jsp?file_id=283698 (accessed on August 19, 2016).
6. U.S. Patent ACT. United States Patent and Trademark Office. [Online] http://www.uspto.gov/patent/laws-regulations-policies-procedures-guidance-and-training (accessed on August 19, 2016).
7. European Patent Office (EPO). The European Patent Convention. [Online] October 7, 1977. https://www.epo.org/law-practice/legal-texts/epc.html (accessed on August 19, 2016).

8. Takenaka T. Chief Judge Rader's Contributions to Comparative Patent Law. *Wash. J.L. Tech. & Arts* 2012, 7: 379–404.
9. Supreme Court of the United States. [Online] October 2009. http://www.supremecourt. gov/opinions/09pdf/08-964.pdf (accessed on August 19, 2016).
10. Directive 98/71/EC of the European Parliament and of the Council of 13 October 1998 on the legal protection of designs. EUR-Lex Access to European Union Law. Access to European Union Law. [Online] 1971. http://eur-lex.europa.eu/legal-content/EN/TXT/? uri=uriserv%3AOJ.L_.1998.289.01.0028.01.ENG (accessed on August 19, 2016).
11. Council Regulation (EC) No 6/2002 of 12 December 2001 on Community designs. European Union. [Online] European Union Intellectual Property Office (EUIPO). http://euipo.europa.eu/en/design/pdf/reg2002_6.pdf (accessed on August 19, 2016).
12. Directive 2008/95/EC of the European Parliament and of the Council of 22 October 2008 to approximate the laws of the Member States relating to trade marks. EUR-Lex. Access to European Union Law. [Online] 2008. http://eur-lex.europa.eu/legal-content/ EN/ALL/?uri=CELEX%3A32008L0095 (accessed on August 19, 2016).
13. World Intellectual Property Organization (WIPO). Washington Treaty on Intellectual Property in Respect of Integrated Circuits. [Online] 1989. http://www.wipo.int/treaties/ en/ip/washington/ (accessed on August 19, 2016).
14. EUR-Lex Access to European Union Law. [Online] 2013. http://eur-lex.europa.eu/ legal-content/EN/TXT/?uri=CELEX:52013PC0813 (accessed on August 19, 2016).
15. World Intellectual Property Office (WIPO). Strasbourg Agreement Concerning the International Patent Classification. [Online] 1971. http://www.wipo.int/treaties/en/ classification/strasbourg/ (accessed on August 19, 2016).
16. Oriel S. Hooking the corporation on the value of intellectual assets. *Intellectual Asset Management* 2010, September–October: 91–95.
17. Horton C. The GE IP agenda. *Intellectual Asset Management* 2011, January–February: 27–29.
18. Khavandkar J, Khavandkar E. *Intellectual Capital: Managing, Development and Measurement Models.* Tehran, Iran: Industrial Research and Training Center of Iran Press, 2009.
19. Davis JL, Harrison SS. *Edison in the Boardroom. How Leading Companies Realize Value from Their Intellectual Assets.* New York: John Wiley & Sons, 2001.
20. European Patent Office (EPO). The European Patent Convention, Rule 71. [Online] https://www.epo.org/law-practice/legal-texts/html/epc/2013/e/r71.html (accessed on August 19, 2016).

Section II

ICT-Based Platforms and Technology Examples

5 Introduction to ICT-Based Service Platforms and Patient Record Systems

Madis Tiik and Indrek Ait

CONTENTS

5.1   INTRODUCTION

To give an overview of the ICT-based service platforms and patient record systems, we have divided the chapter into three sections. First, we introduce some background information to understand why we need health informatics and technology in general, then we give a brief history of how this field has evolved during the last few decades, and finally we introduce the fundamental building blocks that are necessary to create these service platforms. Health care is a difficult field because of numerous reasons. It is more or less different in all of the countries in the world that means it is very hard to give a perfect and precise overview of everything that has happened, is happening, and will happen in this sector. The next part gives an overview of different technologies that ICT-based health care service platforms and patient records comprise. They are introduced to the reader from simple to more complex systems. To implement and use complex systems, one has to have adopted the simple systems. It is not about giving the reader a chronological order but a development order. After getting to know the fundamental systems, electronic services of clinicians and consumers are introduced to the reader. Section 5.4 is about fundamentals and the building blocks that are needed to successfully build health service platforms and patient record systems. There are no shortcuts in our opinion. Of course, nowadays one can develop and adopt the principle structures and systems much faster, but they still need to be developed and adopted starting from the ground up.

5.1.1   TRENDS AND NEED

As described in Chapter 2, there are quite many important trends that affect the health care sector enormously. Health care is very complex and quite inefficient industry. Health care and medicine specifically is an information and image-intensive field. Hospitals need to be efficient and timely in their works and this includes their information management. Information management is the critical part of decision making in medicine as it has to be timely, relevant, and reliable. As the amount of clinical knowledge has become very large, computers are of great help. Machines are able to store, process, manipulate, and retrieve huge amounts of information and images efficiently and quickly. As delivery of health care may have to deal with physical, geographical, and other barriers, the field of informatics can help here a lot to overcome these hurdles. It is important to transmit and share medical and health information across organizations and enterprises using local area networks or wide area networks including the Internet. It is also necessary to provide health care services to remote and underserved areas where regular delivery is too complicated, expensive, or low quality. Technological solutions are crucially needed for sustaining medical education by reaching out to professionals. One of the biggest areas health care is pressured to redesign its thought process is cost-efficiency and here technology is hoped to be the savior. Ultimately going paperless and filmless is meant to reduce costs and make work more efficient. Doctors usually do have not enough time for the visits and there is lack of attention as the number of doctors per patient is too small. Technology also aims to help to provide better care and services. Clinical decision support helps doctors to do better choices. It should be seen clearly now that there is a huge need for health informatics and technologies.

5.1.2 Brief History

Between the 1960s and 1970s, hospitals started to have computers. Together with computers came the first administrative applications. The adoption of these applications was driven by the need to process large volumes of hospital data, which was much easier and better to do with computers rather than by hand. First, the applications were mainly for billing and health insurance claims.

Throughout the 1970s to 1980s, administrative systems started to be replaced by hospital information systems (HISs). During this period, the attempts of integration of clinical components into the administrative and financial systems were starting. Early examples of these information systems include the HELP system that was developed at Latter-Day Saints Hospital in Utah, USA (Collen & Ball, 2015) and the DIOGENE system that was developed in Geneva, Switzerland (Borst et al., 1999).

The 1970s brought along the early developments of digital medical records systems. The aim was to automate the entry, storage, and retrieval of patient data. Examples of these systems are PRoblem-Oriented Medical Information System (PROMIS) from University of Vermont Medical Center (Goldberg, 1988) and TMR from Duke University (Collen & Ball, 2015). PROMIS featured Problem-Oriented Medical Record, which is thought to be one of the earliest medical record systems. It was the first system to link patient and patient care by using hypertext.

During the 1980s, medical informatics was formalized as a scientific discipline. During this decade, professional groups such as the International Medical Informatics Association, the American Medical Informatics Association, and the European Federation for Medical Informatics were formed and medical informatics departments grew in North America, Asia, and Europe.

From 1990 to 2000, technology including Internet was rapidly evolving and gaining mainstream popularity. In medicine, this trend manifested in fast and pervasive adoption of networked and client–server technology. This was the time of growth for Internet and PC-based clinical information systems and knowledge bases. The field of medical informatics development accelerated.

During the change of millennium, several nations started to plan for a country-wide integration platform that would cover all the different health care stakeholders. Denmark, Finland, and Estonia were the first pioneers in this field. Estonia launched its nationwide health information exchange (HIE) platform in January 2009. At the same time, Estonia was the first country to show all collected health data to the patients. For that, a patient portal service was developed, where people could look up health information that was collected about them, give permissions (organ donation permission, prohibition of resuscitation, and blood transfusion), and designate trust-ees and close own health data (OECD, 2010; Sabes-Figuera & Maghiros, 2013).

5.2   SERVICE PLATFORMS AND PATIENT RECORD SYSTEMS

The first step is the transformation from analog to digital, where main element is that hospitals start to gather and use digital data. Historically, it was first mostly used for making insurance and billing tasks easier. Later digital patient data enabled doctors to start communicating better with each other and to do more informed medical deci-sions. Digital health records adoption by health care service providers forms the base and need for integrations. Data exchange systems and document standards have to be in place to move from local networks to broad networks. The central parts of it are integrations and data sharing between different stakeholders. As increasing number of health-related technologies and services are in the hands of the people themselves, patient empowerment is starting to take place, and it is not all about the health care service providers anymore.

5.2.1   Electronic Medical Record

As introduced in Chapter 1, the widely deployed and popular computer applica-tions such as electronic medical record (EMR), sometimes also called as electronic patient record, is in basic version, a digitalized version of the regular traditional paper-based medical chart for each individual. It contains all of the patient's medi-cal and clinical data history created in a single facility, such as a hospital, clinic, or general practitioner (GP) office. EMR includes patient demographics, medical and treatment histories, vital signs, progress and problems notes, laboratory and test results, medications, treatment plans, immunization and allergies information, radiology images, and behavioral and environmental data. Health care providers use it to monitor and manage care delivery within the facility. EMRs have developed a lot over time and nowadays EMR itself comprises a lot of different modules. In our approach of understanding, one example of a modern EMR is an HIS that we will introduce next. The other example of an EMR might be a GP information system. Later we expand on electronic health records (EHRs).

5.2.1.1   Hospital Information System

The main aim of an HIS is to manage the information for health professionals so they can perform their jobs effectively and efficiently. The core system elements are retrieving and submitting patient information, scheduling, admission, discharge, and transfer of patients. The business and financial systems deal with payroll and accounts receivable. Communications and networking systems integrate different parts of HIS and have order entry and result reporting. Departmental management systems focus on the needs of individual departments. Medical documentation is a very important part of the system, where collecting, organizing, storing, presenting, and quality assurance are dealt with.

Hospitals need quite many different systems besides the main information sys-tem to function properly. Many hospitals have as many as 200 disparate systems combined into their HIS, and in some other cases there is only one enormous system used. Next we bring out three very important components that are part of every hospital: laboratory information systems (LISs), radiology information systems (RISs) including picture archiving and communications systems (PACSs), and clinical decision support systems (CDSSs).

5.2.1.1.1   Laboratory Information Systems

A LIS is computer software that enables authorized technicians to enter, process, store, and manage data from different medical tests and processes when a patient is tested at a hospital, clinic, or some other facility. Clinicians can access the results and reports more quickly that leads to faster diagnoses and treatment. The most common elements of LIS are patient check-in management, entry of orders and results, processing of specimens, and patient demographics. Clinical details about patients' laboratory visit are tracked and stored. LIS improves efficiency and lowers costs by lowering the number of unnecessary and duplicate tests.

5.2.1.1.2   RIS and Picture Archiving and Communications Systems

The RIS including diagnostic imaging (DI) systems and supported by PACSs forms the basis for fast and efficient workflow in the radiology departments. The RIS is a networked digital database that is used in the radiology departments of hospitals for collecting, storing, managing, and distributing patient data and images. Its aim is to control and document the work that is done with the patients. The system has patient tracking and scheduling, result reporting, and image tracking functions. Reports and images can be quickly added, retrieved, and transmitted. It allows scheduling appointments and tracking patients through different stages of diagnosis and treatment. DI systems allow health care providers to view diagnostic images such as ultrasounds, MRI and CT scans, and x-ray images. There is no need for a film, as all the images and reports are in electronic format. DI systems are supported by the PACS for digital archiving of these images. The PACS provides economical and efficient solution of digital storage and access to images, interpretations, and related data from different modalities compared to the traditional film-based solutions. Thanks to the RIS and PACS the productivity of health care providers increases. As it reduces turnaround, patients get diagnosed faster and treatments can start sooner.

5.2.1.1.3   Clinical Decision Support Systems

As patient treatment and management decisions in health care are strictly needed to be evidence based and patient centered, clinicians need smart tools to offer best medical services. Health care providers have been increasingly starting to use knowledge management technologies at the point of care to support clinical processes. One important tool is CDSS. These are computer applications that are designed to help health care providers to make better decisions by giving evidence-based case-specific advice, guidelines, reminders, and alerts. These systems divide broadly into two types. The first one uses inference engines to analyze big data sets of known medical knowledge for comparison with a particular patient's information. The second system is without medical knowledge base and is established solely on machine learning algorithms for analyzing clinical data.

By combining the best research evidence with clinical expertise and the specific patient's needs, the level of treatment rises and clinical workflow streamlines. The system is not anticipated to replace clinicians but to play a basic supportive role by making sure that health care professionals can more precisely identify differences of patient conditions and be able to keep track of patient responses to the therapy. A good way to think of it is that the main goal of implementing advanced CDSS is to decrease errors and improve patient safety, improve quality through adoption of best practices, increase cost-effectiveness, and optimize the management of chronic diseases (Scheepers-Hoeks et al., 2009). There are different kinds of CDSSs. Current Medical Diagnosis and Treatment shows the relations between CDSS functional classes (Papadakis et al., 2014):

- *Feedback*: Provide feedback by responding to an action taken by the clinician or to new data entered into the system
- *Data organization*: Organization and presentation of disparate data into logical, intuitive schemas at the point of need
- *Proactive information*: Provision of information to the clinician at the point of need (e.g., clinical pathways on different medical conditions)
- *Intelligent actions*: Automation of routine and repeated tasks for the clinician on a regular time schedule
- *Communication*: Alert clinician and other providers who need to know about unusual data or communications regarding specific patients
- *Expert advice*: Diagnostic and therapeutic advice using a comprehensive knowledge base and a problem-solving method, such as probabilistic reasoning, neural nets, or heuristic rules

5.2.2 ELECTRONIC HEALTH RECORDS

While EMR is the first step forward in patient care, there are problems and limitations between different EMRs interoperability and communication. These systems traditionally cannot understand each other and as there are very many different facilities and systems they use, it is not possible to build integrations between all these connections. Usually, there is not enough incentive for them to do it either. This means that a particular patient's data might be scattered among different facilities, which means that doctors might not always have all the information that there is to get about that patient. This also means that sometimes duplicate tests and examinations have to be performed that is inefficient and unpleasant. One solution to this is an EHR, which is an official real-time digital version of a patient's traditional paper health record or chart that is available and shared instantly and securely among multiple health care facilities within a community, region, state, or in some cases the whole country.

Like EMRs, EHRs are longitudinal patient-centered records that contain a patient's full health profile starting from the first attendance or admission to a health care facility. EHRs surpass standard clinical data collection as they give a broader view of patients' health by getting information from specialists, laboratories, and other health care facilities that are involved in the patient's care. Electronically inserting medical information by health care providers from different facilities over a long period of time allows EHR to evolve into a real useful lifetime record of health information. The aim of EHR is to automate and streamline health care providers' workflow. It is very important to ensure that information generated in EHR is timely, accurate, and available all the time.

Another difference between EMR and EHR is that for EHR information can be created, gathered, managed, and consulted by all the authorized clinicians at the same time at numerous locations and organizations that are involved in a patient's care. Sharing information with various health care organizations and different service providers is built into EHR's core functionality. This enables information from specialists, pharmacies, laboratories, medical imaging facilities, emergency care centers, and other clinics that are involved in the patient's care to be combined together to form a whole picture of the patient's present and past health and medical situation. This means that EHRs contain much more data and give a more comprehensive overview of patients' health data when compared to a single EMR, which contains information created only in a single facility. An EHR makes it possible for the patient record to move together with the patient, unlike an EMR that is strictly bound to a certain facility. EHR like EMR also supports other care-related technologies and activities such as quality management and decision support systems.

EHR is a legal record, created by qualified health care professionals and maintained typically by a health care service provider. In many countries, only the doctor responsible of care can enter data to the EHR. Furthermore, the responsible health service provider has by law the responsibility of the data quality of the EHR. In most countries, there are specific regulations, which define who can access the EHR and how it can be distributed, how long, and by whom it shall be archived.

Governments, industry companies, and academic medical centers have developed different kinds of EHRs. Many nations have different incentives to ensure that EHR covers all the citizens. Over the years, different forms of EHRs have been developed and implemented. Some nationwide EHRs are currently being planned or developed. Types and coverage of EHRs vary a lot, and governments define differently what they exactly mean by EHRs. Basque Country in Spain has solved their data exchange problems with the implementation of a single EHR system to all the providers (Díaz, 2014). North Ireland has a different solution. They integrated the health care record with the social care record to one united system called care record (Health and Social Care Board, 2014). If there is the same system in place in all the facilities in the region, there are no interoperability and data exchange problems as it is based on the same standards. Creating EHRs that work as described before is not an easy process. To ensure successful implementation, some very important elements have to be in place that are discussed in Section 5.4.

5.2.3 HEALTH INFORMATION EXCHANGE

While EHR is a leap forward in patient care, there still persist problems and limitations between different EHRs interoperability. As there is a growing number of EHRs, the need for communication between these systems is crucial for effective large-scale HIEs.

The lack of standards for exchanging patient information between different health care providers means that EHRs that are built by different companies cannot exchange information outside their private networks. The barriers of different systems not understanding each other are holding patient care from moving forward and realizing the true benefits in regard to health information technology usage and possibilities. The next developmental step is going from separate EHRs that cannot change information with each other to using HIE that becomes a bridge between different systems and enables exchange of patient data on the national or regional scale. The purpose of HIE is to securely and reliable transfer health care-related data among diverse systems at numerous facilities, organizations, and government agencies. HIE enables doctors, nurses, pharmacists, and other health care providers to access, retrieve, and share patient data. It aims to improve health care delivery speed, quality, and safety and reduce costs. As data are combined and standardized from various sources, they can seamlessly integrate and give more complete patient records in the hands of the physician and improve care.

HIEs can technically be designed differently, either a series of hub-and-spoke repository systems or a point-to-point information exchange system. The United Kingdom, Norway, Canada, Estonia, Finland, and others chose a hub-and-spoke or similar approach. A point-to-point or similar approach is chosen by New Zealand, Australia, Denmark, and the Netherlands to name a few. The hub-and-spoke system collects information in repositories through which health care providers can access consolidated information. In the point-to-point system, every provider has its own database and shares information from it when asked by a health care provider from another facility. To get a complete overview, the health care provider has to make many requests to various health care providers that hold the patient data (Canada Health Infoway, 2015).

5.2.4 PERSONAL HEALTH RECORDS

Personal health records (PHRs) have evolved from health and training diaries and one-device applications to personal health hub and cloud services and are now moving toward becoming true health accounts where EHR and PHR merge and add an extra layer of analytics that is extended to gene, environmental, and social data. The main difference of PHRs compared to EMRs and EHRs is that they are controlled totally by the person herself or himself. If the PHR makes possible to share data with a health care provider, then it is up to the person to share it or not.

We are first writing about health and training diaries because we like to think of them as early versions of PHRs. Although holding a single purpose, they are digital records of some sort of particular health measurement, activity, or behavior. These systems are standalone or connected to some particular devices. For example, more sophisticated running watches, digital glucose meters, blood pressure monitors, weight scales, or peak flow meters might have their own digital environment for storing training or health data. This data flow might be automatic via wire or wireless connection or manual. The main aim is to store data and get a better overview for more efficient health management. Sometimes it is possible to share this data with friends or with health care workers. More about different kinds of devices, sensors, and applications for producing health, activity, and behavior-related information is introduced in Chapters 1, 6–9 and 12–14.

When we want to understand PHRs, it is necessary to first remember that EHRs are legal records managed by health care professionals. Some people and organizations think that the PHR is just an extension of the legal EHR. Others accept the view that the PHR is a record owned and managed by a person and an organization stores the PHR on behalf of the person and handles the system.

The PHR does not replace any legal record of the health care provider, as there are two separate systems with two different viewpoints on a person's health. So far specific legislation regulation that there is for the EHR does not exist for the PHR. Consumers have numerous options to choose from. By nature PHR is a global record, but there are also some national pilot projects where the aim is to provide every citizen a PHR account.

A more sophisticated PHR of a person usually consists of information collected and integrated from multiple different sources relevant to health, wellness, and care. Any kind of health-related data are possible to be inserted or uploaded manually or automatically from apps and devices. The individual manages and controls this content and can grant permissions for access or share it with others. The aim of the PHR is to improve health. There are different use cases for people. It might be used to make better health-related decisions, to get better overview of one's health-related indicators, and be more engaged and knowledgeable about their level of fitness. It is also possible for a person to manage prevention or monitor chronic disease and health status. Also as people can share this information with a health care provider, its aim is to improve the quality of work of the health care providers as they might get much more data about the individual health behavior and subsequently do better informed decisions. The wide variety of PHR systems in existence gives numerous possibilities for the users. Some PHRs add extra layer of services such as messaging between doctors and patients, reminders, appointments, and more. The scene is rapidly changing as new and more sophisticated systems emerge.

In some cases, PHR can automatically get data from EHR also. This might turn useful to get even a better overview of health. When added an extra layer of information, new services, and data analytic capabilities, this new kind of system might be called for distinguishing purposes a health account to represent a fuller, much smarter, and personalized way to manage a person's health. Sweden and Australia are moving toward this. At the moment of writing the book, Australia has an EHR that is controlled by the patient and soon some real PHR services are planned to be added. Sweden is developing a PHR besides an EHR, but they plan to integrate them. The aim is to combine PHR and EHR data and possibly add even more data sources and build new intelligent services that can provide value from this broad spectrum of data. The person would have the control over the integrated record. EHR would still be for the health care providers but they would send data to health account also.

5.3   SERVICES

In this section, we introduce some of the most important and used digital services for clinicians and for the patients that are integrated with service platforms and health record systems.

5.3.1 Clinician Electronic Services

5.3.1.1 Computerized Provider Order Entry

Computerized physician order entry (CPOE) is a patient management system for authorized health care providers to enter medication, tests, procedures, and treatment orders for the patients under his or her care at the point of care or offsite. Inserted orders are transferred over the network to the departments and medical staff that have to fulfill the order. These can be orders for pharmacy, laboratory, radiology, surgery, and so on. CPOE eliminates the need for handwritten and verbal orders. The aim of the CPOE system is to decrease delays and reduce errors by providing more reliable communication between the medical staff and by ensuring digital, complete, and standardized orders. CPOE systems include standardized order sets that are based on evidence-based clinical guidelines. They are usually connected with CDSSs that typically provide error checking for incorrect tests or doses and duplicates and suggest routes and frequency of drug administration. It might also check for drug allergies or interactions between drugs.

5.3.1.2 Electronic Referrals

Electronic referrals or e-referrals are closely related to virtual consultations. E-referrals are digital versions of paper referrals. E-referrals give the authorized health care provider a way to place new referrals from other health care service providers for their services electronically and not by a paper note, fax, or mail like before. This makes patient care faster and easier for the clinician and the patient. Dependent on the system, it might be possible for the clinicians and patients to track the statuses of their referrals.

5.3.1.3 Virtual Consultation

Virtual consultation or e-consults are secure, direct, and documented consultative communications between two or more health care providers, usually between primary care and specialist care workers. This can be done through a shared EMR, EHR, or HIE. E-consults are closely related to e-referrals when the responding facility's (to whom the e-referral is placed to) provider validates the urgency of the problem, described in the referral and according to judgment, asks the patient to come to the appointment, or gives GP advice how to manage the problem. This type of e-consults avoids unnecessary visits to specialists and is a tool for better communication between GP and specialists. Its aim is to improve access to speciality knowledge without the need of physical meeting.

5.3.1.4 Electronic Prescribing

Electronic prescribing or e-prescribing is the digital version of paper prescribing. E-prescribing gives an authorized health care provider a way to securely generate and renew medical prescriptions electronically and not by a paper note, fax, or mail. This is done through EHR and pharmacy management software, which allows the patient conveniently and safely get the correct medication from the pharmacy of his or her own choosing. Compared to regular script writing prescriptions, it is aimed to decrease errors and adverse drug-related events and increase the overall quality of care.

5.3.2   Consumer Electronic Health Services

5.3.2.1   Remote Patient Monitoring

Remote patient monitoring (RPM) uses technological solutions to enable providing care outside of traditional clinical settings (see health technology in Chapter 1). It connects the patient with health care providers that are located at different locations by collecting medical and other health data from the patient in one location and transferring this data to the health care provider. Then the health care providers can assess this data and give out medical recommendations. RPM features such as physical parameter monitoring and trend analysis give out timely information to the caregivers about the changing conditions of the patient. Then it is possible to take actions before it requires going to the emergency center and stay in the hospital. Chronic disease management is an important part of RPM as people can manage their conditions independently from their own homes with minimal costs without complications and disruptions of visiting a care facility. RPM gives comfort to patients and their families because they know that they are constantly monitored and supported if needed, especially if care involves difficult self-care processes. RPM might have a positive effect by saving time, decreasing costs, and increasing patient satisfaction, quality of life, and access to better care. An example of issues related to stroke self-care is discussed in Chapter 10.

5.3.2.2   Remote Patient Consultation

Remote patient consultation (RPC) or e-visits are secure, documented, and direct consultative communications between health care providers and a patient. RPC can include phone calls, video conferencing, emails, text messages, and so on. This enables new ways of interaction for the patient to conveniently connect with his or her health care providers from wherever the patient or the providers are. RPCs are needed when a patient has a quick question, needs a simple service that usually would not need a physical encounter, or there are issues with distance, mobility, or transportation of the patient if for example they live in remote rural areas far away from health care providers. If systems for RPCs are connected to EMR, EHR, or HIE, the information for the consultation, if needed, can be gathered from these official systems and also entered into the patient record after the consultation. RPCs with greater functionality make possible for the patient to update their vital signs, allergies, and other information. RPCs can increase patient satisfaction by decreasing avoidable physical encounters and reducing time and costs of visiting a doctor. Also it might free up health care providers time.

5.3.2.3   Electronic Booking

Electronic booking or e-booking is the digital version of regular phone or in person appointment scheduling with a health care provider. E-booking has benefits for patients and providers. For the patient, it is more convenient, easier to use, and might include digital reminders, which reduce missing the appointments. Appointments can be made 24/7. For providers, it saves substantial time, improves staff satisfaction, and decreases the number of people who do not show up or cancel in the last minute. As staff spends less time scheduling appointments on phone, it opens up time for more valuable work.

5.3.2.4   Patient Portals

Patient portals (eHealth portals) are online access points that enable patients to use different services, communicate, and interact with health care providers and share medical and health information with each other remotely. Patient portals can be used 24/7. There are different types of patient portals. Some applications are standalone websites, others are integrated into health care providers existing websites, and some are modules added on top of EMR systems. There could be several different patient portals in use for the patient at the same time from different service providers. These patient portals can include and show different information and have different services. This is a real shortcoming, as the information is fragmented and inconvenient to access. The aim of patient portals is to increase efficiency and productivity for patients and health care providers.

Patient portals offer different services. The core functionality of every patient portal is the ability for the patient to securely access health information that has been collected about them by the health care providers. This makes it possible for the patients to understand their health better and take more active stance for improving it. On the other hand, people might want to check that their health information is correct. Next functionality that many patient portals have is RPCs. Patients can communicate with their caregivers by sending messages, commenting, or asking questions. Patient portals might also enable patients to fill some registration forms that reduce visitation time in a health care facility. Electronic booking might be also part of patient portals. In patient portals, people can give permissions such as organ donation permission and prohibition of resuscitation or blood transfusion. People can also designate trustees and close their own health data. Some patient portals also allow to order eyeglasses and contact lenses. Patient portals might also enable online requests for prescription renewals and refills. This way patients do not have to physically visit their health care providers or call them. eHealth portal services can be quite useful for the patients and help patients do better informed decisions and reduce inefficiencies. As can be seen, there are different services available and new ones being constantly developed.

5.4   FUNDAMENTALS

There are some fundamental aspects needed to be covered to successfully implement and use ICT-based service platforms in health. Common IT infrastructure, robust standards to support health IT, use of unique patient identifiers, privacy issues related to health IT systems, access rights, authentication, and consent management are elements that have to be thought of and taken care of to successfully develop, implement, and use these ICT-based service platforms.

5.4.1   COMMON IT INFRASTRUCTURE

Building shared IT infrastructure, that is, technology that can be used by multiple health care providers, helps lower costs and increase interoperability by creating a shared platform for health care organizations to use. Examples of common health IT infrastructure include shared EHR systems, online authentication services, electronic billing systems, secure email, online portals, and health data networks.

5.4.2   ROBUST STANDARDS TO SUPPORT HEALTH IT

Robust standards are critical to the effective application of health IT and play a crucial role in spurring the use of new technology. To facilitate the standard-setting process, many governments actively engage with all stakeholders, including those from the private sector, to coordinate the development of standards. This enables different stakeholders to understand and communicate with each other and it helps to improve compatibility, interoperability, safety, repeatability, and quality, and reduce issues that are connected to them. Standardization is one of the most critical conditions for fluent data exchange between health care organizations. Involving affected professional groups in the early stage of the standardization is the critical factor for later acceptance among those specialties who will start using those documents.

5.4.3   UNIQUE PATIENT IDENTIFIERS AND AUTHENTICATION

Unique patient identifiers help facilitate data sharing between different health care organizations and benefits of their use include reduced risk of medical errors, improved efficiency, and better privacy protection for patients. The use of unique patient identifiers is common in many of the global leaders in health IT, including Denmark, Finland, and Sweden (Protti, 2008). Usage of registries that identify patients and authorized health care providers accurately is essential for electronic records. Registries increase security and reduce administrative time and costs.

Integrated systems need centralized identity administration and user recognition. An identity and authentication solution based on smart card technology currently provides a best of breed foundation for improving EHRs in a secure, private, and sensitive way. Many governments worldwide are currently making efforts for technological modernization and innovation. In Estonia, for example, these secure authentication tools are compulsory ID cards and optional Mobile ID. As another option for secure identification inside hospital settings, to reduce errors when administering drugs to hospitalized patients, hospitals use bar coding or radio frequency identification (RFID) wristbands at medication administration and electronic medication administration records. More about patient identification by RFID can be found in Chapter 8.

5.4.4   PRIVACY ISSUES RELATED TO HEALTH IT SYSTEMS

Although there are policies existing in health care organizations, they are not capable of protecting privacy in electronic health care. It has been said: "Having extensive privacy policies on an enterprise does not directly ensure privacy protection if there is no effective means of consistent policy enforcement across multiple applications and across enterprise boundaries" (Reed, 2005). Enforcement of the privacy policy is essential to ensure that personal information is accessed, used, and disclosed in accordance with ethical norms, so privacy policies should be represented electronically and be managed through software tools capable of detecting the underlying hidden errors. This approach will enable health care organizations to enforce a policy while sharing information with other health care organizations (Mizani & Baykal, 2007).

The key issue when creating databases for collecting and managing the entire population data is building the trust in society and its members. Trust is a two-way relationship that requires constant effort for creating and maintaining it. In addition, trust is extremely relevant to many different levels and relationships existing in health care system, that is, doctor–patient, doctor–doctor, doctor–health care system, and patient–health care system. It is clear that digital health information system will fundamentally change the current doctor–patient relationship and the relationship between doctors and doctors' relationship with the health care system (Sutrop, 2006).

5.4.5　Access Rights

The EHR systems are becoming more and more sophisticated and include nowadays numerous applications, which are not only accessed by medical professionals but also by accounting and administrative personnel. This could represent a problem concerning basic rights such as privacy and confidentiality. Granting access to an EHR should be dependent on the owner of the record; the patient: he or she must be entitled to define who is allowed to access his or her EHRs, besides the access control scheme each healthcare organization may have implemented. In this way, it is not only up to health professionals to decide who have access to what, but the patient himself or herself. Implementing such a policy is walking toward patient empowerment that society should encourage and governments should promote (Falcao-Reis et al., 2008).

In modern, electronic-based healthcare organizations, individuals from diverse professional backgrounds work collaboratively in decision-making processes concerning patients' health, and patients have the right to influence these processes. Winkler states that an organization-wide policy that covers all individuals in a health care organization and deals with both standard and morally controversial medical practices ensures autonomy, quality, fairness, and efficiency of decision-making processes. The privacy policies of many developing countries, which mainly assume a traditional physician–patient decision-making approach, fall short of fulfilling such goals (Winkler, 2005).

5.4.6　Consent Management

Collection and exchange of patient information in EHR creates new threats to privacy and can increase the risk of patient data misuse. Therefore, it is very important to strengthen the patient autonomy with patients' one-time informed consent ("opt-in") or right of patients to limit using their personal data in treatment process ("opt-out"), but also right to add data or make statements of intent. Exchange of patient information should be limited to medical professionals functionally connected to the patient treatment process, for example, concept of attending physician. Access to the data should be equal to all professional groups to support effective treatment processes (e.g., holistic view of the patient does not recognize interdisciplinary limits in the treatment process). Prosecution and punishment functions create an environment of mistrust in the health system. Instead, there is a need to strengthen the patient–health professional relationship by using appropriate requirements for patient counseling and informed consent applications.

Opt-in is the option where the patient's approval is asked to move health data between different service providers and registries. With this the patient designates in what way his or her data can be collected and shared. Experiences of countries such as the UK and Australia show that the number of people who care about data sharing and opt-in is low. Opt-out is the option where data collection and usage rules are set by the government. In such a case, the scope and goals of collection and exchange of health data are established according to the law. If a person does not want to share his or her health data, he or she has to inform the health care facility or submit a corresponding declaration of intent by electronic means. For the wider use of health data and easier creation of new services, the opt-out principle has recently become increasingly more common.

5.5 SUMMARY

This chapter aimed to give an introduction to ICT-based service platform and patient record systems. It gave overview of the trends and needs of health care. It gave a brief history of how ICT-based service platforms and patient record systems were developed and how they evolved over time. An overview of EMRs including HISs together with LISs, RIS, PACSs, and CDSSs was also given. From EMRs, we went to EHRs that are more complicated health record systems that bring together data from various different facilities. Next HIE was introduced, which is the platform that connects information from various different EHRs, EMRs, and other sources to one seamless record and facilitates information flow between providers. Subsequently, going from provider side to personalization patient controlled health diaries, PHRs and health accounts were introduced. Next clinician electronic services such as CPOE, electronic referrals, virtual consultation, and electronic prescribing were discussed. After that consumer electronic health services such as RPM, RPC, electronic booking, and patient portals were discussed. Lastly, the fundamental elements for successful implementation of ICT-based health service platforms and EHRs were covered. Common IT infrastructure, robust standards to support health IT, unique patient identifiers and authentication, privacy issues related to health IT systems, access rights, and consent management are the core elements that need to be covered for successful implementation.

ABBREVIATIONS, TERMS, AND SYMBOLS

CDSS   Clinical decision support System
EHR   Electronic health record
EMR   Electronic medical record
HIE   Health information exchange
HIS   Hospital information system
LIS   Laboratory information system
PACS   Picture archiving and communications system
PHR   Personal health record

RIS   Radiology information system
RPC   Remote patient consultation
RPM   Remote patient monitoring

REFERENCES

Borst F, Appel R, Baud R, Ligier Y, Scherrer JR. 1999. Happy birthday DIOGENE: A hospital information system born 20 years ago. *Int J Med Inform*, 54(3), 157–167.

Canada Health Infoway, Toronto. 2015. *EHR 2015: Advancing Canada's Next Generation of Healthcare*, Toronto, Canada.

Collen MF, Ball MJ. 2015. *The History of Medical Informatics in the United States*. Springer-Verlag, London.

Díaz JMF. 2014. Leveraging eHealth for transforming health services in the Basque Country. Retrieved from: https://www.nll.se/publika/lg/kom/Evenemang/ALEC/Presentations%20 Speakers/Jesus%20Maria%20Fernandez_ALEC%202014.pdf.

Falcao-Reis F, Costa-Pereira A, Correia ME. 2008. Access and privacy rights using web security standards to increase patient empowerment. *Stud Health Technol Inform*, 137, 275–285.

Goldberg A (ed.). 1988. *A History of Personal Workstations*. New York: ACM.

Health and Social Care Board. 2014. eHealth and care strategy for northern Ireland. Retrieved from: http://www.hscboard.hscni.net/consult/Previous%20Consultations/2014-15%20 eHealth_and_Care_Strategy_Consultation/Consultation_Document-e-Health_and_Care-PDF_2mb.pdf.

Mizani MA, Baykal JJ. 2007. A software platform to analyse the ethical issues of electronic patient privacy policy: The S3P example. *J Med Ethics*, 33, 695–698.

OECD. 2010. *Benchmarking ICTs in Health Systems*. OECD, Paris, France.

Papadakis MA, McPhee SJ, Rabow MW (eds.). 2014. *Current Medical Diagnosis & Treatment*. McGraw-Hill Medical, New York.

Protti D. 2008. A comparison of how Canada, England and Denmark are managing their electronic health record journeys. In: *Human, Social and Organizational Aspects of Healthcare Information Systems*, Beuscart R, Hackl W, Nøhr C (eds.), pp. 203–218. Hershey, PA: IGI Press.

Reed A. 2005. What privacy?—*Digital ID World*, 72, 68–70.

Sabes-Figuera R, Maghiros I. 2013. *European Hospital Survey: Benchmarking Deployment of e-Health Services (2012–2013)*. Joint Research Centre of the European Commission Seville, Spain.

Scheepers-Hoeks AM, Grouls RJ, Neef C, Korsten HH. 2009. Strategy for implementation and first results of advanced clinical decision support in hospital pharmacy practice. *Stud Health Technol Inform*, 148, 142–148.

Sutrop M. 2006. Trust and risk in the context of human genetic databases. In: *Ethics: Interdisciplinary Approaches [Eetika: interdistsiplinaarsed lähenemised]*, Simm K (ed.), pp. 243–256. Tallinn: Eesti Keele Sihtasutus.

Winkler EC. 2005. The ethics of policy writing: How should hospitals deal with moral disagreement about controversial medical practices? *J Med Ethics*, 31, 559–566.

6 Data Acquisition, Validation, and Processing in Smart Home Environments

Andrea Kealy and John Loane

CONTENTS

6.1 INTRODUCTION

In this chapter we study the sensor, database, and data processing infrastructure needed to support Ambient Assisted Living (AAL) in smart home environments. AAL refers to unobtrusive systems placed within people's own homes to enable them to live independently. It consists of common sensors such as presence sensors and contact sensors that are found in home alarm systems along with resource usage sensors. Ambient sensors are chosen over body-worn sensors to reduce the impact and stigma of health care devices in the home and deal with issues of poor compliance associated with body-worn sensors. The aim of AAL is to trigger alerts indicating critical events such as a fall and indicate deviations in routine, which may indicate declining health without any input from the user.

We start with an overview of the existing key smart home deployments supporting AAL. Key smart home deployments have to meet the requirement of being deployed in the homes of real people over a long period of time, at least 1 month. This brings unique challenges regarding labeling of data and access, which will be discussed in Section 6.2. There are many more smart home deployments on university campuses that are occupied for short periods of time by research volunteers.

We consider which sensors are most useful and cost effective and what activities of daily living (ADL) can be extracted from sensor data. In a case study of Great Northern Haven, we consider how to validate and monitor the performance of sensors and how to store and process large amounts of sensor data.

6.2 EXISTING KEY SMART HOME DEPLOYMENTS

This section outlines key smart home deployments. The sensor solutions presented in this section have been deployed within the homes of real people over long periods of time.

6.2.1   CASAS

Washington State University's Center for Advanced Studies in Adaptive Systems (CASAS) smart home testbed consists of three disparate setups. These include the CASAS smart home testbed (Kyoto), three single-occupancy apartments within an assisted care facility (Bosch), and the two-story home of two adults and their pet (Cairo) (Rashidi et al., 2011). Kyoto encompasses two setups: the first was home to two adults and integrated 71 sensors (Kyoto1); the second was used to perform scripted activities and integrated 24 sensors (Kyoto2).

Sensors included ceiling mounted motion sensors located 1 m apart; ambient temperature sensors; sensors to record use of the stove, hot and cold water; contact sensors attached to doors and cabinets; and pressure sensors to monitor usage of medicine containers, cooking tools, and the telephone. The Bosch and Cairo setups integrated a combination of motion sensors and contact sensors on doors in key areas, with 32 sensors being installed in the Bosch setup and 27 sensors installed in the Cairo setup. These datasets cover a period of between 2 and 3 months.

The CASAS smart home testbed has been used to generate models of user behavior and recognize activities. Their research focus is on activity recognition and discovery to improve health and energy efficiency. A primary aim relates to performing cognitive assessment (Dawadi et al., 2013) and functional health assessments longitudinally within smart home deployments, in addition to providing assistance in carrying out ADL through prompting, as required (Das et al., 2011; Holder and Cook, 2013). Activity recognition was carried out to identify sleep, bed–toilet transitions, personal hygiene, bathing, meal preparation, eating, leaving the house, taking medicine, cleaning the house, and relaxing or watching TV. Visualizations representing ADL were presented.

Later work focuses on the deployment of a "smart home in a box" system consisting of 24 passive infrared (PIR) motion and light sensors, a door sensor, a relay, and a temperature sensor (Cook et al., 2012). The activities of eighteen single occupancy apartments, integrated with the "smart home in a box" system, were analyzed for a period of 1 month. The aim was to evaluate their proposed approach to activity recognition for different people. Activities assessed included sleep, relax, work, enter the house, hygiene, cook, bed–toilet transitions, clean, leave the house, and medication. Activities were manually annotated and a fivefold cross-validation was used to evaluate the activity recognition labeling, resulting in a weighted average accuracy of 84%. Trends in activity density, sleep quality, time spent in certain areas of the house and health were visualized.

6.2.2   University of Amsterdam

University of Amsterdam's (UvA's) testbed consists of sensors installed in the homes of nine subjects. The sensor setup involves the integration of a commercially available RFM DM 1810 wireless network kit, consisting of a base station and wireless sensor nodes, with analog and digital input sensors (van Kasteren et al., 2008). Sensors installed include PIR sensors, magnetic sensors on doors, cupboards, and the fridge. In addition, a floating sensor may be installed in the cistern of each resident's toilet and a bed pressure mat may be used in the bedroom. Supplementary data from functional health status is being collected via a number of questionnaires, such as the assessment of motor and process skills (AMPS) (Fisher, 1999).

Early work in the UvA focused on activity recognition (van Kasteren et al., 2008). Current research focuses on associating sensor data and activities with functional health status using an ambient sensor setup in the homes of older adults (Robben et al., 2012; Robben and Krose, 2013). Further research focuses on addressing older adults' attitudes to ambient assistive living technologies (Kanis et al., 2012, 2013).

Robben and Krose (2013) describe results of a longitudinal study correlating sensor data with functional health status using an ambient sensor setup. The setup involved ambient sensors installed into the homes of five subjects since summer 2011 and four additional subjects since fall 2012. Sensor data was considered in 1-week blocks in an attempt to deal with outliers. Four objective measures were considered including: raw sensor count, based on the number of firings from all sensors in 1 week; proportional location feature, based on the number of firings of a group of sensors, proportional to the number of sensors within that location; location inference feature, representing time spent in each location; and transition feature set, representing the number of transitions between locations based on the location inference feature. Global correlations indicated that more activity in the bedroom, denoted by the proportional number of sensor firings in the bedroom, was also linked to a lower functional health status. Lower time outside the home was linked to a lower functional health status. Furthermore, a higher number of transitions was associated with a lower functional health status; this may well result from the lower duration spent outside the home. A high variability was identified between individuals when functional health was analyzed on an individual basis. Finally, the research in UvA indicated that individuals with low functional health status tended to isolate themselves. Further research within UvA provided visualizations of long-term deviations in behavior (Robben et al., 2013).

Despite the authors highlighting the diversity between individuals, further research in UvA relates to obtaining better global features for their functional health assessment system. The introduction of 3-monthly functional health assessments adds context to medium-term changes in physical health status.

6.2.3 TIGERPLACE

The University of Missouri's TigerPlace is employed as a test facility and incorporates three different neighborhoods. The first two neighborhoods, consisting of 32 and 23 apartments, respectively, house older people who would have otherwise made the move to a nursing home. The third neighborhood comprises five skilled nursing units, each housing 17 residents, two units are intended for long-term care, one is a nursing home and the other is for patients suffering from dementia or Alzheimer's disease, the other three units are focused on rapid rehabilitation. On-site support is provided from health care professionals, such as nurses and physical therapists. Maintenance, housekeeping, transportation, and dining services are also provided and social activities are scheduled regularly.

An in-home monitoring system within each of the homes in TigerPlace integrates a combination of ambient sensors including PIRs to capture movement in a room as well as localized activity involving use of the fridge, kitchen cabinets, the front door and shower; a pneumatic bed sensor to detect bed restlessness; temperature sensors to identify stove and oven activity; and, in a subset of residences, cameras or Microsoft Kinects (Alwan et al., 2003). Motion detection from PIRs was investigated and represented using visualizations that display all motion sensor events within a single visualization (Wang and Skubic, 2008). The PIRs, used in this setup, transmit an event every 7 seconds when movement is detected. PIRs have also been used to assess dissimilarity of movement levels between individuals as well as temporally to assess changes in an individual's behavior to infer when a change in health may have occurred (Wang et al., 2012). PIRs have also been used to estimate time away from home (Wang and Skubic, 2008) and visitor times (Wang et al., 2011).

Work ongoing at the University of Missouri (Rantz et al., 2005; Wang and Skubic, 2008; Wang et al., 2011) focuses on the development of an early warning system to determine, and consequently alert, when a change in normal behavior occurs. Their research aims to monitor and assess potential problems in mobility and cognition of older people in their homes; to detect falls and changes in daily patterns that may indicate onset of a health problem (Rantz et al., 2008). The setup in Missouri combines expertise from researchers and health care professionals to develop an electronic health record system designed to assist physicians in assessing changes in health, emerging from objective measures, and reducing nursing workloads (Popescu et al., 2011). The goal of the deployment in TigerPlace was to develop and evaluate interventions, and the associated technology, to facilitate aging well while aging in place.

Early research investigated the use of silhouettes, extracted from video, adopting hidden Markov models for temporal pattern recognition of activities based on features extracted from the silhouettes (Anderson et al., 2006). This method identifies when an individual is walking, kneeling, or falling. Further research involved the use of multiple cameras to label the type of fall (Anderson et al., 2009). Features were extracted from a three-dimensional approximation of an individual, or voxel person, and fuzzy inference was used to determine the state of the individual. Recent research focuses on the use of the Xbox Kinect in the longitudinal measurement of gait (Stone and Skubic, 2011a, 2011b).

6.2.4 Oregon Center for Aging & Technology (ORCATECH)

Oregon Health and Science University's Oregon Center for Aging & Technology (ORCATECH) consists of a Point of Care Laboratory, designed to simulate an apartment and 265 homes of community-dwelling older adult participants taking part in the intelligent systems for assessment of aging changes (ISAAC) study (Kaye, 2010). New technologies are tested in the laboratory before being deployed to participant's homes.

Unobtrusive monitoring technologies, including motion and contact sensors, were installed in each home. Each participant was equipped with a desktop computer, broadband, and was given computer training.

Research ongoing in ORCATECH relates to validating the potential of technologies within the home to identify behavior change associated with the early symptoms of cognitive decline (Kaye et al., 2011). Research includes the longitudinal analysis of activity, gait, outings, nighttime behavior, and human computer interaction. Each participant of the ISAAC study was required to be aged 80+ (or 70+ for non-Whites and individuals living with a participant aged 80+) and living independently in a home with more than one room.

ORCATECH gathers weekly subjective reports relating to medication changes, falls, injuries, health changes, emergency room visits, depression, changes in living space, vacations, and visitors (Kaye et al., 2011). Such subjective reporting adds context in analyzing events.

6.2.5 GREAT NORTHERN HAVEN

The Great Northern Haven (GNH) is a purpose-built demonstration housing complex in Dundalk, Ireland. It consists of 16 independent living apartments each of which has been integrated with a series of ambient sensors and actuators intended to monitor and support its residents. Fifteen of the apartments are permanently occupied by older adults. The apartments are real peoples' homes, and as such, the data being collected is extremely rich. The older adults living in GNH are not research volunteers, as is the case with most research projects, providing a unique setup. The 16th apartment is designated as a transition unit for falls prevention. As this apartment is only occupied for short periods, it is available for testing when vacant.

The 15 permanently occupied apartments accommodate 16 older adults, 11 men and 5 women, aged 60–88 (72.8 ± 8.2 years). One apartment is home to a married couple, making it multiple occupancy, while the remaining apartments are occupied singly. The first residents moved into GNH in June 2010 and sensors have been gathering data since June 23, 2010. The residents suffer from a wide range of illnesses including heart conditions, diabetes, depression, and bipolar disorder, with one resident having previously suffered a stroke and another having suffered from cancer. Each resident of GNH lives there permanently; hence their apartment in GNH is their home. As such they represent a diverse cohort of real people living independently in their own homes.

There are a total of 2240 sensors and actuators throughout GNH, with approximately 100 installed in each home. Figure 6.1 illustrates the sensor layout within each apartment. Each apartment has been fitted with:

- *PIR sensors*: Three PIR sensors detect motion and interior brightness levels in the living room, hallway, and main bedroom.
- *Contact sensors*: On all windows and exterior doors as well as on three interior doors between the hallway and living room, the hallway and main bedroom, and the main bedroom and en suite.
- *Light switch sensors*: On all light switches.
- *Temperature sensors*: In each room.
- *Electricity and heating consumption sensors*: Detecting power consumption and heating usage for each apartment.

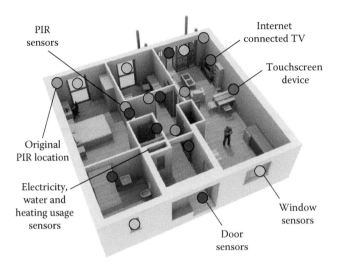

FIGURE 6.1   GNH sensor layout.

A subset of actuators are also installed in each resident's apartment. These include:

- Alarm cords and buttons, a home security system, and a telecare device that links with a monitoring service.
- Optional actuators, installed in a subset of apartments, include Picomed hydraulic actuators on doors and actuators to raise and lower the height of the kitchen sink and cooker hobs.

Aside from the sensors and actuators installed within each apartment, contact sensors have been installed on the external doors used to access apartments on the first and second floors. A number of sensors located on the roof aggregate weather-related data. These include outside brightness, temperature, wind speed, and rainfall, saved every 10 minutes. Cameras have also been installed in front of the GNH complex and within the communal entrances. Thirty days of archive footage is stored; however, these are used for security purposes only.

The research at GNH focuses on assessing the physical and emotional well-being of the residents from the sensor data (Doyle et al., 2014). Movement levels, time outside the home, location within the home, sleep metrics, electricity usage, and weather are all used as indicators of these two domains. These metrics can be combined with subjective data, gathered from the residents, to assess their physical and emotional well-being.

Subjective data relating to physical and emotional well-being are gathered using clinically validated questionnaires. Questionnaires include the Center for Epidemiologic Studies Depression scale (CES-D) (Radloff, 1977), the Hospital Anxiety and Depression scale (HADS) (Zigmond and Snaith, 1983), the Montreal Cognitive Assessment (MOCA) (Nasreddine et al., 2005), the 36-item short-form health survey (SF-36) Quality of Life questionnaire (Ware Jr and Sherbourne, 1992), the De Jong scale of emotional and social loneliness (de Jong-Gierveld and Kamphuls, 1985), the Instrumental Activities of Daily Living (IADL) scale (Lawton and Brody, 1969), and the Pittsburgh Sleep Quality Index (PSQI) (Buysse et al., 1989). These questionnaires provide quantitative baseline health measures for each resident.

YourWellness is an iPad application that allows individualized questions to be asked of residents to understand their repeated patterns of behavior as well as gathering their own report on their well-being. Residents can see their trends in sleep, mood, and social interaction over time and can annotate associated graphs to help label the data.

Data gathered from GNH is processed and fed directly back to the residents. The aim is to encourage behavior change in the resident through awareness of their own wellness rather than sharing the data with health care professionals or relatives.

6.3   SENSORS TO SUPPORT AAL

Having presented the key smart home deployments above, we now consider what sensors are most useful to support AAL. Many of these deployments are experimental in nature and have installed a large number and variety of sensors so it is interesting to consider which sensors are most useful and cost effective. Table 6.1 summarizes the sensors installed in each of the key deployments.

As can be seen from Table 6.1, there is no standardization of sensors to support AAL in smart homes. Motion sensors and contact sensors are the only two

TABLE 6.1
List of Sensors in Each of the Key Smart Home Deployments

	CASAS	UvA	TigerPlace	ORCATECH	GNH
Motion	✓	✓	✓	✓	✓
Contact sensors	✓	✓	✓	✓	✓
Bed sensors		✓	✓	✓	✓
Gait sensors			✓	✓	
Phone sensors	✓		✓	✓	
Vital signs sensors			✓		✓
Food	✓				
Cooking	✓		✓		
Medication	✓		✓	✓	
Lighting					✓
Water usage	✓	✓			✓
Electricity usage	✓			✓	✓
Heating					✓
Temperature	✓				✓
Audio			✓		
Video			✓		

sensors that have been installed in all of the deployments. These sensors are cheap and familiar from home security systems and give useful information on movement levels within the home as well as time spent outside the home. When combined, the sensors can be used to map where residents are in their home at different times of the day.

Bed sensors have been installed in all but one of the deployments. These sensors are more expensive and less familiar than simple motion and contact sensors.

Medication, phone, water, and electricity sensors have been installed in three of the deployments. Medication and phone sensing are more invasive than other sensors considered so far but gather important information on medication adherence and social interaction. There is an important balance to be reached between gathering data and respecting privacy. Water and electricity sensors give important information on activities such as showering and cooking. These sensors are more expensive to buy and often require expert installation, which is an added cost.

Gait sensors, vital signs sensors, cooking sensors, and temperature sensors have been installed in two locations. Gait sensors such as Xbox Kinect or Shimmer and vital signs sensors such as Withings weighing scale or blood pressure monitor give useful data but require action by the participant. This can lead to issues such as poor adherence. Cooking can be inferred from electricity or gas usage sensors, so a specific sensor to detect cooking is unnecessary. Temperature sensors are cheap and give useful information on comfort levels so it is a surprise that they have not been deployed more widely.

Audio, video, food, lighting, and heating have been installed in only one deployment. Audio and video are generally not acceptable to participants due to privacy concerns. Food sensing is difficult to do due to the number of foods and ingredients. Lighting sensors that detect light levels are perhaps unnecessary when contact sensors on light switches can tell whether lights are on or off. Heating is the main energy cost in northern European countries but is often difficult and expensive to monitor due to the variety of heating systems.

6.4   BEHAVIOR MEASURES

Having considered the sensors installed in each of the key smart home deployments we now consider which behavior measures each tries to extract. As noted with the sensor setups, there is no standard set of sensors and equally there is no standard set of measures extracted from the data gathered at each site. This is a relatively new area of research and is not yet mature enough to have established standards. Also, the heterogeneity in the abilities of the participants being monitored means that a one size fits all solution is unlikely.

Table 6.2 shows the behavior measures being extracted by each of the key smart home deployments. Researchers have indicated that an increased volume of walking is highly beneficial for functionality (Fortes et al., 2013). Physical activity levels tend to decline as people age (Murtagh et al., 2014). Retaining physical function is fundamental to maintaining well-being as people age. Activity as measured by movement levels and time outside the house are common at all sites. These two measures are extracted from movement sensors and contact sensors on external doors of the

TABLE 6.2
List of Behavior Measures Examined by Each of the Key Smart Home Deployments

		CASAS	UvA	TigerPlace	ORCATECH	GNH
Activity	Movement levels	✓	✓	✓	✓	✓
	Room transitions		✓		✓	✓
	Time outside the home	✓	✓	✓	✓	✓
	Visitors/multiple occupancy			✓	✓	
Sleep	Total time in bed			✓	✓	✓
	Bed time and rise time				✓	✓
	Classification of heartbeat and respiration			✓		
	Bed restlessness			✓	✓	✓
	Bed-to-toilet transitions/number of bed exits	✓			✓	✓
Activities of daily living (ADL)	Personal hygiene	✓				
	Meal preparation	✓		✓		
	Eating	✓				
	Taking medicine	✓				
	Cleaning the house	✓				
	Relaxing	✓				
	Computer use	✓			✓	

homes. Room transitions and detection of visitors are subject to the sensor setup in the homes. Room transitions are easily detected if there are movement sensors in each room with full visibility of the room while detection of visitors requires more advanced algorithms and additional sensors (Petersen et al, 2012).

Degradation in sleep over time has been shown to have a considerable negative impact on health (Happe, 2003; Stanley, 2005). Subjective estimates of sleep appear to differ from objective measures in certain populations (McCall et al., 1995; Buysse et al., 2010). The number of awakenings has been identified to be a measure which individuals have difficultly identifying subjectively (Argyropoulos et al., 2003). It has been suggested that it is important to measure both acute and longer term trends due to the high variation in sleep patterns (Hayes et al., 2008). Sleep and sleep measures have been extracted at four of the sites. There are two distinct approaches taken toward gathering the sleep measures. One involves low-cost movement and contact sensors along with location mapping algorithms. Large periods of time where the participant is in the bedroom, the lights are off, and movement levels drop below a threshold are classified as sleep. The second approach is to use more expensive bed sensors such as Emfit that can monitor heartbeat and respiration or wrist-worn acti-graphs such as Withings Activite Pop. Again adherence with wearing sensors, cost and privacy are potential issues with this approach.

None of the other ADL have wide use across deployments. To extract these ADLs accurately, it is necessary to add extra sensors or interact with the participant to label data. Hoque and Stankovic (2012) have applied data mining techniques to cluster sensor firings so that each cluster represents instances of the same activity. This way users only need to label each cluster as an activity and the system can recognize future similar activity.

6.5   CASE STUDY: DATA GATHERING, SENSOR VALIDATION, AND DATA PROCESSING IN GREAT NORTHERN HAVEN

Details of the sensors and behavior measures extracted from the data at GNH have been given in Section 6.4. Here, we will consider the low-level details of the data-bases, data validation, and data processing that are carried out to archive data and extract behavior measures from the sensor data gathered at GNH.

6.5.1   Network and Databases

All sensors installed in GNH are connected using a wired Konnex (KNX) network. Sensor data is collected on a dedicated server in GNH, logged by NETxAutomation OPC server software and stored in a local Microsoft SQL Server 2008 r2 Express edition database. All sensor firings are also written to log files, stored locally, once a day. Data is replicated from the local database to a production MySQL server in CASALA via a virtual private network every 15 seconds.

All data is initially replicated to a daily table before being processed and inserted into a monthly table. This ensures that the daily table remains small. Hence inserts into both the daily and monthly tables are quick. Processing converts sensor identi-fiers that have been stored as varchars, comprising the eib_address (varchar(20)), ip_address (varchar(30)), and phys_address (varchar(20)) to identifiers stored as smallint, including sensor_id (smallint(3)) and apartment_id (smallint(3)). As the size of the database is reduced, the monthly tables fit in memory resulting in the queries executing much faster, as no disk swapping is required.

6.5.2   Data Loss

Even though sensors in GNH are connected via a wired KNX network, there are many issues that can cause data loss. It is useful to have a dashboard as in Figure 6.2, where at a glance you can see that you are getting expected number of sensor read-ings. This dashboard can help to identify when there are issues with the sensors, power outages, or internet connection issues.

A script runs every 30 minutes to check that data is being stored on the local SQL Server database in GNH. If the timestamp of the latest entry in this local database was recorded more than an hour previously, an email is sent to the team indicating

FIGURE 6.2   Sensor readings dashboard.

that all systems should be checked. If the internet connection in GNH is interrupted, the replication script automatically copies all unreplicated data once the connection is restored.

To date data loss has resulted from a number of issues. The first was caused by the use of Microsoft SQL Server 2008 r2 Express edition in GNH. When the database exceeded the size limit of 4 GB, data failed to be written to the local database and consequently this data was lost. This has been solved by removing records more than 120 days old from this database, as long as they have been replicated to the production server. As a secondary solution log files are written to the local server in GNH, with a new log file created once a day. This ensures that in the event that data is not saved to the local database it may be copied from the log files.

Data loss was also experienced when the Windows-based boxes in GNH shut down as a result of Windows updates and power outages. Windows updates have been disabled to avoid data loss where possible but power outages remain an issue.

One major instance of data loss resulted from a hardware failure. Gateways for data aggregation lost their IP addresses. While the failure was identified quickly the expertise required a third party to reinstate IP addresses. Hence 2 weeks' worth of data was lost, between June 1, 2012, and June 13, 2012.

Quality checking measures, considered in more detail below in the Data Validation section, have been implemented to ensure that sensor failures are identified. Malfunctions in individual sensors require a third-party installer to be called, which results in a delay in fixing these sensors. To date, the failure of individual sensors has affected two movement sensors and one contact sensor on an external door.

6.5.3   SENSOR VALIDATION: PIR VISIBILITY

A method, similar to that described in Kaushik et al. (2007), was used to validate PIRs within the test apartment. The validation was carried out in each room containing a PIR, as well as adjacent rooms within line of sight of each PIR. These areas included the open plan living room and kitchen; the hallway, as well as entrances to the toilet and guest bedroom; and finally the main bedroom and en suite.

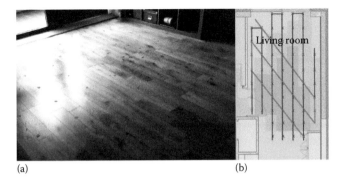

(a)   (b)

FIGURE 6.3   GNH floor validation markers (a) and path of floor validation (b) within the living room.

Each validation area was subdivided into 14-inch blocks parallel to the doorway above which the PIR is mounted. Each point was marked with tape as illustrated in Figure 6.3a. Continuous motion, from the waist up, was carried out on each point for the overshoot time of 10 seconds plus an additional 5 seconds. This additional time facilitated the differentiation between persistent movement on each point and transitions to each point. This method was initiated on each point parallel to the PIR bearing wall, following the red lines in Figure 6.3b, and also diagonally, along the blue lines in Figure 6.3b. The use of multiple path directions facilitates the validation of movement toward and away from each PIR.

The resulting PIR visibility for the test apartment in GNH is shown in Figure 6.4. Each white circle indicates a ceiling mounted PIR. Wall-mounted PIRs are represented as white semicircles. The standard three PIR setup, applicable to all GNH apartments, is illustrated in Figure 6.4a. Areas with full PIR visibility are colored dark

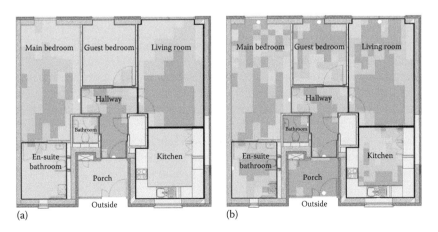

(a)   (b)

FIGURE 6.4   Locations overlaid on standard PIR visibility in GNH apartments (a) and visibility with additional PIR sensors (b).

gray. If a PIR did not register a value of 0 while moving on a point, it was determined to be fully visible to the PIR. In the three-PIR setup, this accounted for 32.727% of the living room, 42.857% of the hallway, and 22.857% of the main bedroom.

Areas that detect transitions to them but do not detect subtle movement within them are colored mid gray. If a PIR registered a value of 1 when moving to a point but registered a value of 0, 10 seconds after moving to this point it was deemed to be partially visible to the PIR. In the three-PIR setup, this accounted for 30.090% of the living room, 46.429% of the hallway, and 37.143% of the main bedroom. Areas undetectable to PIRs are colored light gray. If a PIR does not register a value of 1 when movement is initiated to a point, it is considered to have no PIR visibility. These areas include the row closest to the wall bearing a wall-mounted PIR, the corner by the rear door and the majority of rooms that do not contain PIRs. In the three-PIR setup, this accounted for 36.363% of the living room, 10.714% of the hallway, and 40% of the main bedroom.

A second evaluation was carried out in the test apartment subsequent to the installation of additional PIR sensors on August 31, 2012. The additional PIR setup, illustrated in Figure 6.4b, achieves a higher level of PIR visibility. The additional PIRs were only installed in the test apartment and are not extended to other GNH apartments.

6.5.3.1 Sensor Validation: In Isolation

Each sensor has been validated in isolation to ensure that the underlying sensor and data collection framework is functioning as expected. A number of tests are conducted daily for each individual sensor to identify periods in which sensors have experienced downtime or failure. The number, frequency, and consistency of sensor firings, as well as the typical range of values, are evaluated to identify whether an individual sensor is faulty or has failed. Additional checks relate to sensor type, comprising "periodic," "on change," and a combination of "on change" and "periodic." For instance, when evaluating an "on change" sensor, such as a PIR, the time between a PIR triggering and resetting is checked to ensure that it is greater than the overshoot time. The evaluation criteria for each sensor type are considered next.

"Periodic" sensor validation, illustrated in Figure 6.5a, is the most straightforward due to periodic sensors reporting a constant number of records each day.

If the number of records output by a periodic sensor is below the expected range, it demonstrates that this sensor has experienced data loss. For instance, internal/external brightness sensors should fire every 5/10 minutes. Hence, if less than 288/144 firings have been registered in a given day, this sensor has experienced data loss.

"On change" sensor validation, illustrated in Figure 6.5b, considers event-driven sensors that report only when a sensor changes in value. In GNH "on change" sensors include PIRs, temperature sensors, light switches, and the wind speed sensor. These sensors are validated by looking at a sliding window of the previous 30-day period to assess whether the number of firings falls within typical ranges. Figure 6.6 illustrates the variation in the cumulative number of light switch sensor firings within apartment 3 each month. Increased sensor firings are evident following the resident's move to GNH in November 2010. The 30-day sliding window prevents

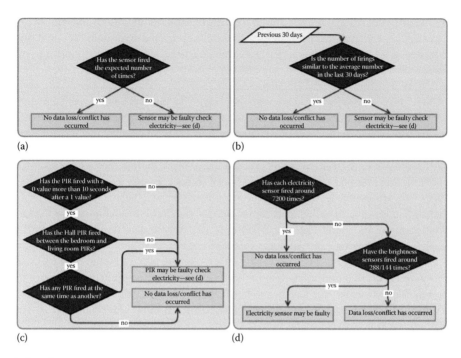

FIGURE 6.5   Smoke testing of periodic sensors (a); on change sensors (b); and PIRs (c) as well as detection of data loss (d).

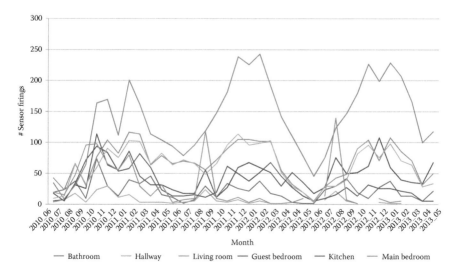

FIGURE 6.6   Cumulative monthly sensor firings for each light switch sensor in apartment 3.

erroneously detecting changes caused by natural seasonal variation. The number of light switches sensor firings, specifically the living room, main bedroom, and outside, appears to decrease below normal levels in June 2012. This dip in sensor usage was confirmed to coincide with an extended period of sensor downtime experienced across all apartments in the GNH complex.

The third sensor type, encompassing both "on change" and "periodic" sensor behavior, may be validated in the same way as periodic sensors, illustrated in Figure 6.5a. In GNH, this sensor category includes contact sensors, installed on windows and doors, which fire every 15 minutes as well as each time they are opened or closed. Hence, each of these sensors should fire at least 96 times per day. In addition to the aforementioned quality checking faulty contact sensors may register as being "open" regardless of the door or window's status. Hence, a rule is applied to verify that contact sensors installed on external doors should report as being "closed" more often than "open." This rule is applied only to external doors due to the security implication in leaving external doors open.

Additional validation is applied to PIR sensors, as indicated in Figure 6.5c. PIR sensors issue an end of movement output value of 0 exactly 10 seconds after the last detected movement. Hence, our first step in validating PIRs is to verify that each has fired with a value of 1 when movement is detected, changing to a 0 value 10 seconds after motion has stopped. If no movement has been detected by a PIR for an extended period, a value of 0 may be output to indicate that the absence of movement does not result from sensor downtime. Hence, PIRs should not register a value of 1 more often than a value of 0.

6.5.3.2   Sensor Validation: Consistency between Sensors

Aside from checking PIRs in isolation, individual sensors may be verified against each other. Sensors employed to infer a resident's location, using the home movement mapping model, should not fire in different rooms concurrently unless visitors are present. It follows that no two PIRs should pick up the same movement unless the visibility of PIR overlaps. Hence, no two sensors should fire at the same time in a single occupancy apartment. Furthermore, transitions between rooms should be detected. If, for example, movement ceases in the living room, we would expect the PIR in the hallway to detect movement before the PIR in the bedroom. These two assumptions highlight the need to validate the visibility of PIRs to verify: (1) if the movement of one individual may be detected by two PIRs simultaneously; and (2) if room transitions are detected.

In addition to checking that sensors in disparate locations do not trigger concurrently, data loss may be detected by checking high-frequency periodic sensors. For instance, if data loss has occurred for an individual sensor, a secondary validation, depicted in Figure 6.5d, should be completed. This verifies whether the individual sensor is faulty or whether downtime has occurred within the apartment. As the electricity sensor, installed in each apartment, reports most frequently (every 12 seconds), it is used to verify that data has not been lost from any apartment on any given day. Establishing that each electricity sensor has reported approximately 7200 records each day, that is, that five electricity firings were recorded each minute ($5 \times 60 \times 24 = 7200$), provides a strong indication that downtime has not occurred. If the number of records is considerably lower, it indicates that data loss has occurred within this apartment.

6.5.4   DATA PROCESSING

Once data has been processed and partitioned into monthly tables on the production server, the data is then processed further using scripts run as cronjobs to extract metrics such as movement, location, and sleep for each apartment for the previous day. Each script writes the processed data to summary tables in the production database. We now discuss these metrics.

6.5.5   MODELS AND METRICS

Figure 6.7 illustrates the approach applied to derive the behavior measures, integrating the following three models:

1. PIR-derived movement model, inferred from raw sensor data, used to measure:
   a. Total movement (TM)
2. Home movement mapping model, inferred from raw sensor data, used to measure:
   a. Time outside the home (TOTH)
   b. Total movement inside the home (TMITH)
   c. Number of transitions (#Trans)
3. Sleep and Sleep-disturbance measurement model, inferred from the PIR-derived movement model and the home movement mapping model, used to infer:
   a. Bed time (BT)
   b. Rise time (RT)
   c. Time in bed (TIB)
   d. Total movement during sleep (TMS)
   e. Number of disturbances (#Dist)
   f. Duration of disturbances (DurDist)
   g. Number of night-time transitions (#NtTrans)

6.5.5.1   PIR-Derived Movement Model

The PIR-derived movement model uses raw PIR sensor data to infer residents' movement levels within a predefined time period.

The PIR-derived movement model measures movement within the standard three PIR setup in GNH. Each bout of movement is calculated as the time between a PIR firing with a value of 1 and the time the PIR resets to a 0 value.

Movement levels are extracted individually for each PIR, in each apartment, and segmented into 1 minute, 1 hour, and 1 day epochs. The 1-minute movement epochs are accessed by the sleep and sleep-disturbance measurement model. The hourly movement has been visualized for each PIR (O'Brien et al., 2012) to identify patterns in each of the three living areas, specifically the living room, hallway, and main bedroom, over time. Hourly movement, inferred from PIRs within each

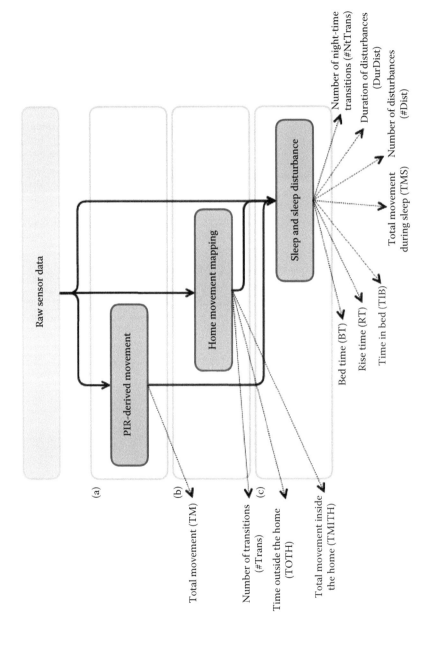

FIGURE 6.7 Framework to model movement (a); room location (b); and sleep and sleep disturbance (c).

home, has been examined in Wang and Skubic (2008), Shuang et al. (2009), Rantz et al. (2010), and Wang et al. (2011, 2012). Finally, daily movement incorporates all movement detected from midnight on 1 day until midnight the next.

6.5.5.2 Home Movement Mapping Model

The home movement mapping model is used to infer an individual's location within their home throughout the day, hence may be used to calculate the time spent in each room as well as the time spent outside the home. For the purpose of this discussion, areas are divided into two categories:

1. Rooms containing PIRs, known as *PIR rooms.*
2. Areas that do not have a PIR installed, including the kitchen, guest bedroom, WC, en suite, and outside. In GNH, all rooms without PIRs are located adjacent to a PIR room; hence these are known as *adjacent locations.*

The home movement mapping model extracts each resident's location from raw sensor data. The location of PIRs, above the entrance to the hallway, living room and main bedroom, can detect room transitions. Hence, it is possible to definitively state that an individual has entered a PIR room when the corresponding PIR fires with a value of 1. Movement to adjacent rooms is calculated using a subset of sensors, including windows, doors and light switches, installed within each adjacent room. This model provides a determination of the room in which a resident is located at any specific time. Figure 6.8 illustrates how transitions to PIR rooms (a), outside (b), and to adjacent rooms (c) are determined.

6.5.5.3 Sleep and Sleep-Disturbance Measurement Model

Various sleep measures have been derived by considering elongated periods within the bedroom deduced by the home movement mapping model.

The purpose of the sleep and sleep-disturbance measurement model is to estimate the sleep window, illustrated in Figure 6.9a, that is, the period during which sleep is possible. This estimated sleep window is then used to infer sleep metrics including estimated bed time and rise time, illustrated in Figure 6.9b, sleep disturbances and total time in bed, illustrated in Figure 6.9c.

Bedroom and out of bedroom events, inferred using the home movement mapping model, are examined to determine the estimated sleep window, denoted by the dashed box in Figure 6.10. The following questions are asked in choosing the sleep window: "Is the resident located in the bedroom?" "Is the start time between 5 pm and 12 pm? and "Is this grouped in the same sleep window as the last?" These time points facilitate the detection of early nights, beginning in the early evening, and sleep-ins, starting before 12 pm.

An elongated period detected after 12 pm signifies that the sleep window for the previous night has ended, and also marks the beginning of the estimated sleep window for the next night. Sleep data is often examined over nonoverlapping time windows, or epochs. For the purpose of the sleep and sleep-disturbance measurement model, the PIR-derived movement within the

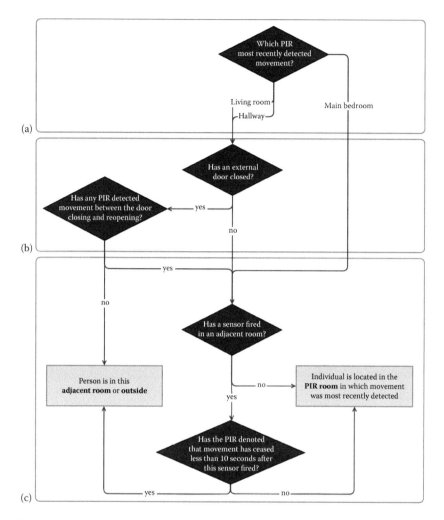

FIGURE 6.8 Home movement mapping model to detect movement to the PIR room (a); outside (b); or adjacent room (c).

bedroom, segmented into 1-minute epochs, has been used. This derived movement data is used to calculate bed time, rise time, and disturbances, discussed in Sections 6.5.6 through 6.5.8.

6.5.6 BED TIME AND RISE TIME

BT is determined as the start of a period where movement is below the movement threshold, set to 3 seconds of movement per minute, for at least 10 epochs. This movement threshold was set as the default threshold for each resident. The movement threshold of 3 seconds was deemed to eliminate periods of light movement occurring

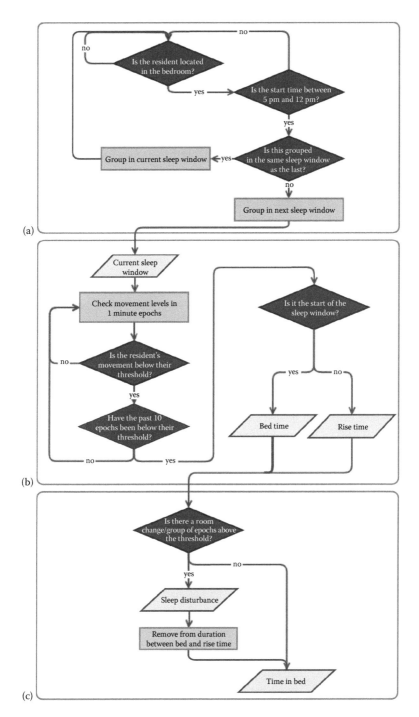

FIGURE 6.9 Sleep window (a), calculating bed time and rise time (b), and identifying sleep disturbances (c) decision trees of the sleep and sleep-disturbance model.

12 am   10 am

FIGURE 6.10 Sleep window estimated from a resident's location map. Includes movement levels (overlaid in black).

during sleep. Actions such as reading in the bedroom may cause movement to dip below the threshold; hence an additional restriction is in place, indicating that the resident is assumed not to be asleep if the bedroom light is on.

RT is estimated by deducing the last period at the end of the sleep window whereby movement is below the threshold for at least 10 minutes.

6.5.7 NIGHTTIME DISTURBANCE MEASURES

Transitions to other rooms, as well as movement detected by the PIR in the bedroom, between the estimated bed time and rise time are considered to be nighttime disturbances. Examples of these disturbances are denoted by the dashed boxes in Figure 6.11.

The first disturbance measure relates to periods surrounding room transitions, indicated by the first dashed box in Figure 6.11. Where a room transition has been detected the assumptions for determining bed time and rise time, detailed in the Section 6.5.6, are applied to calculate the estimated wake time before the disturbance and the estimated sleep time after the disturbance. The total disturbance time is calculated as the duration, in hours, between the estimated wake time and sleep time.

Sleep-disturbance measures include the number of disturbances and the cumulative duration of disturbances recorded per night.

6.5.8 TIME IN BED

The time in bed is inferred from the measures calculated thus far, including bed time, rise time, and disturbance duration. The time in bed is determined to be the duration, in hours, between bed time and rise time, removing periods that were calculated to be disturbances, as indicated in Figure 6.12.

12 am   10 am

FIGURE 6.11 Disturbances and movement detected during sleep, calculated from increased movement levels between bed time and rise time.

12 am   10 am

FIGURE 6.12   Time in bed, that is, time between bed time and rise time eliminating sleep disturbances.

6.5.9 LIMITATIONS

The reliability of the sleep and sleep-disturbance measurement model detects sleep solely within the bedroom; hence sleep initiated within other rooms is not detected by this model.

The model is also limited by the location of PIRs within the bedroom. If PIRs do not have coverage of the bed, they may not fully detect nighttime movement levels, resulting in an earlier bed time, a later rise time, less movement detected during sleep, and as a consequence a longer time in bed.

6.5.10 DAY TIME BEHAVIOR

Day time behavior measures comprise the TM, the time outside the home, the total movement proportionate to the time the resident is at home, and the number of transitions.

6.5.11 TOTAL MOVEMENT

The TM is inferred from the PIR-derived movement model that takes raw PIR sensor data as input. The TM, measured in hours formatted as hh:mm:ss, encompasses movement levels detected by all PIRs cumulatively each day.

6.5.12 TIME OUTSIDE THE HOME

The time outside the home (TOTH) consists of the duration of time, measured in hours formatted as hh:mm:ss, a resident spends outside his/her home. The TOTH is inferred from the home movement mapping model alone. Again, the TOTH each day encompasses the 24-hour period between midnight on 1 day and midnight the next.

6.5.13 TM PROPORTIONATE TO TIME INSIDE THE HOME

The TM proportionate to time inside the home (TMITH) consists of the duration each resident spends moving relative to the time spent at home. It is calculated from derived behavior measures, comprising the TM and the TOTH. The number of hours spent at home (TITH) is calculated by subtracting the TOTH from the full 24-hour period. The TMITH is then calculated as the TM divided by the TITH.

The detected movement is considered proportionate to the TITH each day, hence is output as a percentage. This calculation is comparable to that calculated in Wang and Skubic (2008), Shuang et al. (2009), Rantz et al. (2010), and Wang et al. (2011, 2012), which calculates movement levels based on time inside the home alone.

6.5.14   NUMBER OF TRANSITIONS

The number of transitions consists of the cumulative number of times a resident was determined to move to a different room each day. The number of transitions is again inferred from the home movement mapping model alone. Furthermore, the number of transitions each day encompasses the 24-hour period between midnight on 1 day and midnight the next.

6.5.15   LESSONS LEARNED

In this section, we highlight lessons learned during 5 years of work at GNH. The aim here is to capture practical advice that will help researchers to avoid trial and error in the setup of smart environments to support AAL.

6.5.15.1   Plan for System Maintenance and Use a Dashboard to Monitor Sensors

In a long-term smart home environment, individual sensors fail and the whole system suffers downtime for many different reasons. It is crucial to be aware of this and have a monitoring system in place to limit this downtime. This can take the form of a dashboard that displays system status at a glance and automated emails that detect system downtime and notify the responsible engineer.

Even when a sensor failure is detected, it is important to recognize that access to the participant's home is limited and it is wise to schedule system maintenance with the participants when setting up the system.

6.5.15.2   Validate Sensor Setup

Sensors need to be validated before they can be used to build algorithms. It is important to do this to be aware of uncertainty, which should be captured in the model. PIRs should be positioned to capture room transitions and ensure that you can detect movement in all areas of the home. This may mean having to install more than one PIR if the room is large.

We had an instance where a new PIR was installed in a small bathroom area and on testing, it was not detecting movement. On investigation, it was found that the sensor had been installed high in a wall and aimed at the roof. It was possible to open the sensor and aim it down at the floor so that it detected movement.

6.5.15.3   Have Onsite Database to Cache Data When Connections Are Down

Internet connections fail and firewall rules can be changed and cause communication issues so it is very useful to have a local database at the monitoring site to avoid losing data. Once these connection issues are corrected data can be gathered from the local database.

6.5.15.4   Understand Database Limitations

When gathering large amounts of data every day, it is important to realize that tables and databases will grow quickly and we need to design the system to deal with these issues. Tables can be kept small by partitioning them into daily and monthly tables. This way tables fit in memory and database queries are fast.

Free versions of databases often have size restrictions that can cause data loss once that limit is exceeded. It is important to be aware of this and replicate and delete data, if necessary. It is also useful to plan enough disk space and be easily able to extend disk space as the system grows.

6.5.15.5   Appreciate That Labeling Data Is Difficult

It is relatively easy to gather data from sensors. The difficult part is labeling the data to understand what different activities look like. There are a number of options. You could ask the residents to annotate their activities in their homes, but this is invasive and has poor compliance and is difficult to do for long periods of time. A nicer solution is to cluster the sensor data to find commonly occurring blocks of sensor data and then ask the participants to label just those activities. A final option is to have an application that allows interaction with participants where you can push questions to them or where they can annotate data.

6.5.15.6   Use Off the Shelf Wireless Sensors for
Smart Home in a Box Solution

GNH was built in 2008 and 2009 and has a fully wired KNX network installed. Sensor technologies have moved on a lot since that time and the CASAS group already has a smart home in a Box project. With off the shelf wireless sensors such as SmartThings and OpenEnergyMonitor it is now possible to set up a smart sensing environment for a fraction of the price that the network in GNH cost.

6.6   CONCLUSION

In this chapter, we have presented the main smart home deployments that support AAL. We looked at the common sensors deployed and ADL extracted from the sensor data. It is important to note the heterogeneity in sensor setups and activities extracted and that this is a relatively new research area where common standards have yet to be reached.

We then presented a case study focusing on sensors to support AAL at GNH. The need to validate the sensor setup and monitor sensor and system status was presented. Sensors that report periodically are easiest to monitor and this should be considered when setting up a smart home deployment.

Next we presented a movement model, a location mapping model, and a sleep and sleep-disturbance model that allow the extraction of movement and sleep metrics.

Finally, we presented some practical lessons learned at the GNH deployment. Of particular interest here is how the smart home and sensor market has changed in the last 5 years to be at a point where we can now buy off the shelf sensors and install them in a home using an existing wireless network to create a low-cost smart environment to support AAL.

ACKNOWLEDGMENTS

We would like to acknowledge the team at NetwellCASALA, all of whom have contributed to ongoing research there, including Rodd Bond, Andrew MacFarlane, Benjamin Knapp, Julie Doyle, Ed Lenox, Lorcan Walsh, Brian O'Mullane, and Carl Flynn.

REFERENCES

Alwan, M., Kell, S., Dalal, S., Turner, B., Mack, D. & Felder, R. 2003. In-home monitoring system and objective ADL assessment: Validation study. In: *International Conference on Independence, Aging and Disability*, Washington DC, p. 161.

Anderson, D., Keller, J. M., Skubic, M., Chen, X. & He, Z. 2006. Recognizing falls from silhouettes. In: *IEEE Engineering in Medicine and Biology Society (EMBC)*. IEEE, New York, pp. 6388–6391.

Anderson, D., Luke, R. H., Keller, J. M., Skubic, M., Rantz, M. & Aud, M. 2009. Linguistic summarization of video for fall detection using voxel person and fuzzy logic. *Computer Vision and Image Understanding*, 113(1):80–89.

Argyropoulos, S. V., Hicks, J. A., Nash, J. R., Bell, C. J., Rich, A. S., Nutt, D. J. & Wilson, S. J. 2003. Correlation of subjective and objective sleep measurements at different stages of the treatment of depression. *Psychiatry Research*, 120(2):179–190.

Buysse, D. J., Cheng, Y., Germain, A., Moul, D. E., Franzen, P. L., Fletcher, M. & Monk, T. H. 2010. Night-to-night sleep variability in older adults with and without chronic insomnia. *Sleep Medicine*, 11(1):56–64.

Buysse, D. J., Reynolds III, C. F., Monk, T. H., Berman, S. R. & Kupfer, D. J. 1989. The Pittsburgh sleep quality index: A new instrument for psychiatric practice and research. *Psychiatry Research*, 28(2):193–213.

Cook, D., Crandall, A., Thomas, B. & Krishnan, N. 2012. CASAS: A smart home in a box. *Computer*, 46(7):62–69.

Das, B., Chen, C., Seelye, A. M. & Cook, D. J. (eds.) 2011. *An Automated Prompting System for Smart Environments*. Springer: Berlin, Germany.

Dawadi, P. N., Cook, D. J. & Schmitter-Edgecombe, M. 2013. Automated cognitive health assessment using smart home smart monitoring of complex tasks. *IEEE Transactions on Systems, Man, and Cybernetics: Systems*, 43(6):1302–1313.

De Jong-Gierveld, J. & Kamphuls, F. 1985. The development of a Rasch-type loneliness scale. *Applied Psychological Measurement*, 9(3):289–299.

Doyle, J. et al. 2014. An integrated home-based self-management system to support the wellbeing of older adults. *Journal of Ambient Intelligence and Smart Environments*, 6(4):359–383.

Fisher, A. G. 1999. *Assessment of Motor and Process Skills. Vol. 2, User Manual*. Three Star Press, Inc: Fort Collins, CO.

Fortes, C., Mastroeni, S., Sperati, A., Pacifici, R., Zuccaro, P., Francesco, F., Agabiti, N., Piras, G. & Ebrahim, S. 2013. Walking four times weekly for at least 15min is associated with longevity in a Cohort of very elderly people. *Maturitas*, 74(3):246–251.

Happe, S. 2003. Excessive daytime sleepiness and sleep disturbances in patients with neurological diseases. *Drugs*, 63(24):2725–2737.

Hayes, T., Pavel, M. & Kaye, J. (eds.) 2008. An approach for deriving continuous health assessment indicators from in-home sensor data. Amsterdam, the Netherlands. IOS Press: Amsterdam, Netherlands.

Holder, L. & Cook, D. 2013. Automated activity-aware prompting for activity initiation. *Gerontechnology*, 11(4):534–544.

Hoque, E. & Stankovic J. 2012. AALO: Activity recognition in smart homes using active learning in the presence of overlapped activities. In: 6th International Conference on Pervasive Computing Technologies for Healthcare (PervasiveHealth) and Workshops. IEEE, San Diego, CA, pp. 139–146.

Kanis, M., Robben, S., Hagen, J., Bimmerman, A., Wagelaar, N. & Krose, B. 2013. Sensor monitoring in the home: Giving voice to elderly people. In: *Pervasive Computing Technologies for Healthcare (PervasiveHealth)*. IEEE, Venice, Italy, pp. 97–100.

Kanis, M., Robben, S. & Kröse, B. 2012. Miniature play: Using an interactive dollhouse to demonstrate ambient interactions in the home. In: *Proceedings of DIS 2012*, Newcastle, UK, pp. 11–15.

Kaushik, A., Lovell, N. & Celler, B. 2007. Evaluation of PIR detector characteristics for monitoring occupancy patterns of elderly people living alone at home. In: *IEEE Engineering in Medicine and Biology Society (EMBS)*. IEEE, Lyon, France, pp. 3802–3805.

Kaye, J. 2010. Overview of the intelligent systems for assessment of aging changes (ISAAC) study. *Gerontechnology*, 9(2):122.

Kaye, J. A., Maxwell, S. A., Mattek, N., Hayes, T. L., Dodge, H., Pavel, M., Jimison, H. B., Wild, K., Boise, L. & Zitzelberger, T. A. 2011. Intelligent systems for assessing aging changes: Home-based, unobtrusive, and continuous assessment of aging. *The Journals of Gerontology Series B: Psychological Sciences and Social Sciences*, 66(Suppl. 1):i180–i190.

Lawton, M. P. & Brody, E. M. 1969. Assessment of older people: Self-maintaining and instrumental activities of daily living. *Gerontologist*, 9(3):179–186. http://www.strokecenter.org/wp-content/uploads/2011/08/lawton_IADL_Scale.pdf (accessed on August 19, 2016).

McCall, W., Turpin, E., Reboussin, D., Edinger, J. D. & Haponik, E. F. 1995. Subjective estimates of sleep differ from polysomnographic measurements in obstructive sleep apnea patients. *Sleep*, 18(8):646–650.

Murtagh, E., Murphy, M., Murphy, N., Woods, C. & Lane, A. 2014. Physical activity, ageing and health, Research paper, Centre for Ageing Research and Development in Ireland (CARDI), Belfast, Ireland. http://www.cardi.ie/sites/default/files/publications/Physical%20activity,%20ageing%20and%20health.pdf (accessed on August 19, 2016).

Nasreddine, Z. S., Phillips, N. A., Bédirian, V., Charbonneau, S., Whitehead, V., Collin, I., Cummings, J. L. & Chertkow, H. 2005. The Montreal Cognitive Assessment, MoCA: A brief screening tool for mild cognitive impairment. *Journal of the American Geriatrics Society*, 53(4):695–699.

O'Brien, A., McDaid, K., Loane, J., Doyle, J. & O'Mullane, B. 2012. Visualisation of movement of older adults within their homes based on PIR sensor data. In: *Pervasive Computing Technologies for Healthcare (PervasiveHealth)*. San Diego, CA: IEEE, 252–259.

Petersen, J., Larimer, N., Kaye, J. A., Pavel, M. & Hayes, T. L. 2012. SVM to detect the presence of visitors in a smart home environment. In: *IEEE Engineering in Medicine and Biology Society (EMBC)*. IEEE, San Diego, CA, pp. 5850–5853.

Popescu, M., Chronis, G., Ohol, R., Skubic, M. & Rantz, M. 2011. An eldercare electronic health record system for predictive health assessment. In: *e-Health Networking Applications and Services (Healthcom)*. IEEE, Columbia, MO, pp. 193–196.

Radloff, L. S. 1977. The CES-D scale A self-report depression scale for research in the general population. *Applied Psychological Measurement*, 1(3):385–401.

Rantz, M. J., Marek, K. D., Aud, M. A., Johnson, R. A., Otto, D. & Porter, R. 2005. TigerPlace: A new future for older adults. *Journal of Nursing Care Quality*, 20(1):1–4.

Rantz, M. J., Porter, R. T., Cheshier, D., Otto, D., Servey III, C. H., Johnson, R. A., Aud, M., Skubic, M., Tyrer, H. & He, Z. 2008. TigerPlace, a state-academic-private project to revolutionize traditional long-term care. *Journal of Housing for the Elderly*, 22(1–2):66–85.

Rantz, M. J., Skubic, M., Alexander, G., Popescu, M., Aud, M. A., Wakefield, B. J., Koopman, R. J. & Miller, S. J. 2010. Developing a comprehensive electronic health record to enhance nursing care coordination, use of technology, and research. *Journal of GerontoloGical nursing*, 36(1), pp. 13–17.

Rashidi, P., Cook, D. J., Holder, L. B. & Schmitter-Edgecombe, M. 2011. Discovering activities to recognize and track in a smart environment. *IEEE Transactions on Knowledge and Data Engineering*, 23(4):527–539.

Robben, S., Boot, M., Kanis, M. & Krose, B. 2013. Identifying and visualizing relevant deviations in longitudinal sensor patterns for care professionals. In: *Pervasive Computing Technologies for Healthcare (PervasiveHealth)*. IEEE, Venice, Italy, pp. 416–419.

Robben, S., Englebienne, G., Pol, M. & Kröse, B. 2012. How is grandma doing? Predicting functional health status from binary ambient sensor data. In: *AAAI Fall Symposium: Artificial Intelligence for Gerontechnology*. Arlington, VA.

Robben, S. & Krose, B. 2013. Longitudinal residential ambient monitoring: Correlating sensor data to functional health status. In: *Pervasive Computing Technologies for Healthcare (PervasiveHealth)*. IEEE, Venice, Italy, pp. 244–247.

Shuang, W., Skubic, M. & Yingnan, Z. 2009. Activity density map dis-similarity comparison for eldercare monitoring. In: *IEEE Engineering in Medicine and Biology Society (EMBC)*, IEEE, Minneapolis, MN, pp. 7232–7235.

Stanley, N. 2005. The physiology of sleep and the impact of ageing. *European Urology Supplements*, 3(6):17–23.

Stone, E. & Skubic, M. 2011a. Evaluation of an inexpensive depth camera for in-home gait assessment. *Journal of Ambient Intelligence and Smart Environments*, 3(4):349–361.

Stone, E. E. & Skubic, M. 2011b. Passive in-home measurement of stride-to-stride gait variability comparing vision and Kinect sensing. In: *IEEE Engineering in Medicine and Biology Society*. IEEE, Boston, MA, pp. 6491–6494.

Van Kasteren, T., Noulas, A., Englebienne, G. & Kröse, B. 2008. Accurate activity recognition in a home setting. In: *Proceedings of the 10th International Conference on Ubiquitous Computing*. Seoul, Korea: ACM.

Wang, S. & Skubic, M. 2008. Density map visualization from motion sensors for monitoring activity level. In: *International Conference (IET)*. IET, Seattle, WA, pp. 1–8.

Wang, S., Skubic, M. & Zhu, Y. 2012. Activity density map visualization and dissimilarity comparison for eldercare monitoring. *IEEE Transactions on Information Technology in Biomedicine*, 16(4):607–614.

Wang, S., Skubic, M., Zhu, Y. & Galambos, C. 2011. Using passive sensing to estimate relative energy expenditure for eldercare monitoring. In: *Workshop on Smart Environments to Enhance Health Care*. Seattle, WA.

Ware JR, J. E. & Sherbourne, C. D. 1992. The MOS 36-item short-form health survey (SF-36): I. Conceptual framework and item selection. *Medical Care*, 30(6):473–483.

Zigmond, A. S. & Snaith, R. P. 1983. The hospital anxiety and depression scale. *Acta Psychiatrica Scandinavica*, 67(6):361–370.

7 Machine Vision in Smart Health and Social Care

Mirka Leino and Pauli Valo

CONTENTS

7.1 INTRODUCTION

The development of machine vision technologies has been remarkable in the past 10 years. The applications, and therefore, the research, have mostly been intended for industrial use. Today, the costs and usability of the machine vision equipment have improved, making machine vision applications much easier to employ in everyday use. There are significant advantages in machine vision compared with traditional methods of diagnostics or physical therapy. First, machine vision is a non invasive technology. For instance, there is no need for needles in imaging. It is possible to image wide areas simultaneously by machine vision. This makes the procedure quick and repeatable, even if there is no certainty where to image. Machine vision can be used to monitor health and well-being, for example, the physical and physiological changes in a human body. Because the applications have been designed for industry, new innovative adaptations are needed to make them applicable for enhancing well-being (Leino et al. 2015).

7.2 BASICS OF MACHINE VISION

This chapter describes the basics of machine vision in such a way that a nonautomation expert is able to understand the technology. The basics are summarized by taking into consideration the most important aspects for smart health and social care. The explanations are illustrated by figures and pictures.

Machine vision is often defined as a mechanical sense-imitating human vision. The following definition is a generally accepted definition for machine vision:

> The automatic acquisition of images by noncontact means and their analysis to obtain desired data for use in controlling a process or an activity (Automated Vision Inc. 2006).

Machine vision is an intelligent combination of digital cameras imaging the target, optics transmitting the light to the sensor, lighting systems illuminating the target, and software analyzing the image. Earlier machine vision equipment was mainly intended for manufacturing and other industrial applications, but today, it is spreading into smart health and social care applications, for example, medical applications, personal security applications, and activating and motivating (exercising) applications.

7.2.1 TRADITIONAL MACHINE VISION

A traditional machine vision system (Figure 7.1) consists of:

- Camera
- Optics
- Lighting
- Interface
- Computer
- Analyzing software

7.2.1.1 Choosing a Camera

The camera is the most important part of a machine vision system, and it should be chosen to always include the needed information of the target. Yet, when is the image good enough? First of all, the features of interest must be well defined and contrast must be high, with adequate number of details. Second, the images must

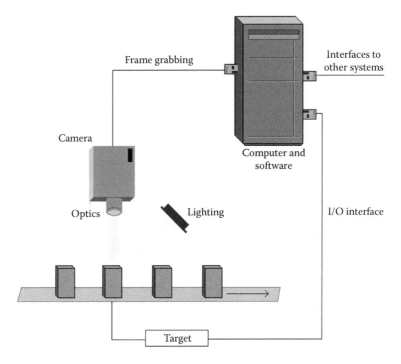

FIGURE 7.1 An overview of a machine vision system.

be repeatable and all the features in the image must also exist in the physical world. In other words, there should not be any noise or ghost images. Third, the possible changes in the environment should have minimal impact on the image.

When choosing the camera and optics for a traditional machine vision system, one should know the target surrounding and lighting. There are a large number of possible cameras, so the selection should be done based on the circumstances, for example, whether the targets are moving or still or how large the target parts are and how far they are.

The type of the camera is chosen when all the factors affecting the image quality have been taken into consideration. The selection depends on whether the system needs a camera with a matrix detector or with line-scan detector. The detector is the part of a camera that receives the light reflecting from the target and creates the image. If the detector is matrix shaped (Figure 7.2), the image will be two-dimensional, and if the detector is line shaped (Figure 7.3), the image will also be a line. A matrix detector is by far the most used detector type in machine vision. It is a good choice when imaging separate or immobile targets.

The line detector is good for imaging moving or continuous-flow-type targets. The line detector is used like a scanner to join the imaged lines together in the computer. The result looks like a continuous two-dimensional image. Figure 7.4 shows the difference between imaging with a matrix camera and with a line scan camera.

FIGURE 7.2   Matrix detector.

FIGURE 7.3   Line detector.

The next choice is done between color and monochrome cameras. According to the general rule, if you do not need to recognize colors, you should always choose a monochrome camera. Monochrome cameras are more accurate because of the estimations made in typical color imaging techniques. Especially if the machine vision system should measure anything, the camera type should be monochrome.

After choosing the camera type, resolution must be determined. Resolution refers to the amount of pixels in the image. In machine vision systems, resolution is expressed in pixels in width and height. There is usually no unambiguity in the determination of resolution. A principal rule is that there should be enough pixels but not too many. There should be enough pixels to see the target, but more the number of pixels, the longer it will take for analyzing, because more number of pixels means that there are more points to analyze. Unnecessary pixels also make the camera more expensive. Some general rules of thumb may be set for pixel resolution requirements. If the purpose of the system is to observe the presence or absence of an object, there should be at least three to four pixels in the object area. To read, for example, a data

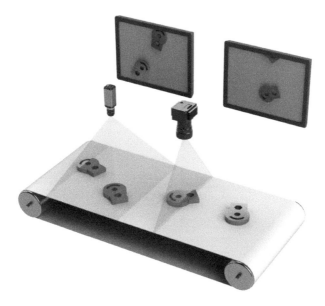

FIGURE 7.4 Matrix camera (on the left) images in two dimensions and line camera (on the right) images one line at a time, and computer combines the lines as a continuous 2D image, as seen on the right.

matrix code, there should be 8–10 pixels per each data matrix cell. In identification of an object shape, there should be from hundreds to thousands of pixels, depending on the complexity of the object shape. In measuring, for example, some geometrical characteristics of an object, the pixel resolution should be 1/10 to 1/3 of the tolerance requirement; otherwise, there should be 3–10 pixels per tolerance value. For example, if the goal is to measure the height of an object with a tolerance requirement of 0.1 mm, there should be at least three pixels per millimeter.

7.2.1.2   Choosing the Optics

When the camera features have been determined, the appropriate optics are chosen. The four main constructs of choosing the optics are (1) focal length (f), (2) working distance, (3) field of view, and (4) detector size (Figure 7.5).

Focal length is a feature of the optics, which is calculated based on working distance, field of view, and detector size. The longer the focal length, the narrower the angle of view from which the camera images (Figure 7.6). There are many focal length calculators for machine vision purposes on the Internet.

Working distance means the distance from the target to the front edge of the camera optics. The field of view is the size of the image area at the focal point. For example, if the object can be seen in an area whose size is 1000 mm wide and 800 mm high and the aspect ratio of the camera is 4:3, the wideness is the critical direction and the field of view must be determined as 800 mm. The size of the object is meaningful only if the object is always seen in the same position. Otherwise, the

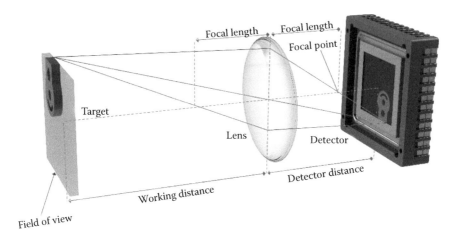

FIGURE 7.5   Significant features in choosing optics.

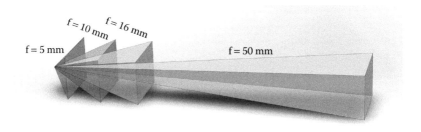

FIGURE 7.6   Comparison of different focal lengths and angles of view.

field of view must be determined as the whole area where the object could be positioned. The detector size is a feature of the camera; therefore, it must be taken into consideration when selecting the camera. The smaller the detector, the smaller the field of view with the same optics.

7.2.1.3   Choosing the Lighting

One of the most crucial phases of designing a machine vision system is choosing or actually designing correct lighting. Its importance cannot be exaggerated when the aim is to implement a stable and timely machine vision system. Good lighting design generates conditions for successful imaging, minimizes the need for image processing, and ensures that the features of interest are seen clearly in the image. If lighting has not been correctly designed, it could be impossible to implement even a simple system.

Lighting design always begins with analyzing the environment of use. When light and shadows, as well as their variations, are known, the next phase is to design and test the lighting of the system. This phase is based on target need recognition, lighting geometry, and technique testing; choosing the lighting source; filtering testing; and testing the interaction of the camera, target object, and lighting.

7.2.1.3.1 Lighting Geometry

When designing lighting, especially the lighting geometry must be considered. Almost any light sources can be used in machine vision. All of them have both good and bad features. The best result is detected by testing. Useful lighting sources are for example:

- LED (light-emitting diode)
- Halogen
- Fluorescent light
- Ultraviolet light
- Metal halide light
- Xenon light
- Infrared (IR) light

Lighting geometry is more important than the lighting source. Lighting geometry refers to the direction of lighting in relation to the target and camera. The most typical lighting geometries have been presented in the following subsections.

7.2.1.3.1.1 Bright Field Bright field lighting illuminates the target from roughly the same direction as the camera images. Bright field lighting is a good choice for general lighting, but it often causes problems in terms of so-called mirror reflections. When the light beam hits the target, the most powerful beam reflects directly to the camera. This results in bright, white areas in the image and prevents the analysis.

7.2.1.3.1.2 Dark Field Dark field lighting (Figure 7.7) brings the light to the target in a small angle in relation to the target surface. Dark field lighting is a good choice with flat, reflective surfaces, if bright field lighting causes mirror reflections.

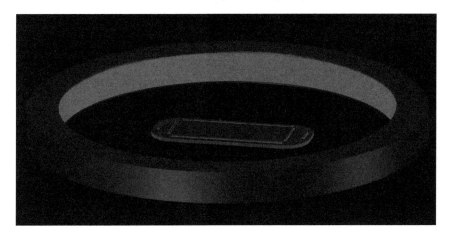

FIGURE 7.7   Dark field ring light.

FIGURE 7.8 Light beams (orange) coming to the target and reflecting (green) from the target.

Dark field lighting is, in particular, used with flat surfaces, when the variations of surface quality should be visible. When the light beam hits the surface from the side, the light reflects to the other side from flat surfaces, while it reflects to the camera from holes, pits, and bulges (Figure 7.8). These surface changes are seen in the image as light areas; however, they are not any lighter on the surface. Respectively, the flat surfaces are detected as dark areas. This feature can be exploited, for example, in analyzing engravings or surface shapes.

7.2.1.3.1.3 Back Lighting   Back lighting is a type of illumination where the subject is between the camera and the lighting source (Figure 7.9). Back lighting is usually used in order to cause the surface of the object reflect as little as possible. The camera sees the object as dark and the background as light. A dark object against a light background creates a perfect silhouette image, which is very suitable for recognizing the outlines of the object. Back lighting is often used in dimension measuring or in other outline inspection.

7.2.1.3.1.4 Diffuse Dome Lighting   Diffuse light means light that comes from more than one point. Diffuse dome light (Figure 7.10) can be produced, for example,

FIGURE 7.9 Camera lighting setup with red back lighting and target object.

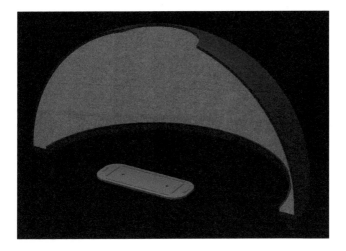

FIGURE 7.10 Cross-sectional drawing of a diffuse dome light.

by placing LEDs on an aluminum ring. The ring is attached to a dome, from whose inner surface, the light beams can reflect to different directions. This dome is placed above the target subject to make the light beams reflect from the inner surface to all possible directions. In this way, the image has no shadows or reflections, and the target is imaged with smooth and equal lighting. There must also be a hole on the upper surface of the dome, so that the camera can image through it. Diffuse dome should be used, for example, with shiny, convex, and concave subjects or with transparent packaging materials, with which other lighting techniques cause too bright reflections.

7.2.1.3.1.5 Axial Diffuse Lighting Axial diffuse lighting (Figure 7.11) is often implemented as an LED matrix, which creates a flat lighting surface. Light beams from this lighting surface are directed to a beam splitter, which directs them to the target object in 45°. In this way, all the light beams travel straight to the target object. The camera images above the axial diffuse light through the beam splitter, and the light beams reflect straight to the camera from flat surfaces, whereas from rough shapes, they reflect to the edges and not to the camera. The flat surfaces are detected as light in the image, whereas the rough shapes are seen as dark. It is an opposite of dark field lighting. In addition, axial diffuse lighting is used, for example, in engraving inspection.

7.2.1.4 Choosing the Analyzing Software

Basically, there are two types of analyzing software for machine vision in the market. The first type is a general-use software for computers. The software can be used with the products of several camera providers, and it usually comes with diverse features and extensive options. This type of software is mostly used with traditional PC-based machine vision systems. The other type of software is

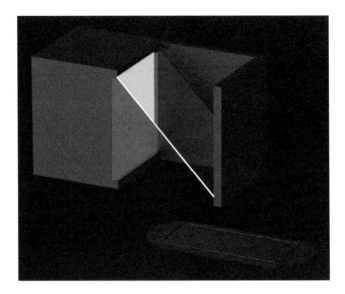

FIGURE 7.11   Cross-sectional drawing of an axial diffuse light.

camera-bound software, which can be used only with a certain type of camera. Analyzing software for smart cameras is usually this type of software. This type of analyzing software often comes with the camera, while general analyzing software usually costs from hundreds to thousands of Euros. In most cases, the analyzing software is chosen on the basis of the experience of the designer.

7.2.2   SMART CAMERA SYSTEMS

Smart camera is an independent camera unit with an integrated processor and memory. While a smart camera images the target and analyzes it without a computer, a traditional PC-based machine vision system uses a camera to image the target and then sends the image to the computer, which analyzes it. A good smart camera can be very quick, although it can never be as quick as a modern computer. However, a smart camera can solve simple problems very efficiently and reliably.

The increase in the use of smart cameras in machine vision has been exceptional. Although smart cameras provide a potential for easier use of machine vision and a more compact setup than the conventional PC-based machine vision, smart cameras also set some limits on the versatility and complexity of the applications. In smart health and social care, smart cameras provide invisible and easy-to-use technology to the applications. Smart cameras are very useful, for example, in inspecting if the patient is falling, if a person with impaired memory has left a coffee maker on, and if there are any security problems in the services for the aged or the disabled. There is rarely a need to send the image anywhere in smart camera applications. As a result, the privacy of the user is not compromised.

7.2.3 3D IMAGING

3D imaging refers to technologies that form 3D models of real items by using different kinds of imaging techniques. 3D imaging serves more information, with new possibilities to identify the target objects and especially their shapes. 3D imaging has developed a lot over the past few years. In smart health and social care, 3D imaging is used, for example, in physiotherapy to instruct the exercises or in surgery to guide the surgeon through the operation.

In most cases, 3D imaging is based on structural lighting, that is, the target is illuminated with some kind of a light structure, such as a laser line or a light dot pattern. There are also other kinds of 3D imaging techniques. This chapter introduces the basics of several 3D imaging techniques.

7.2.3.1 Laser Scanning

Laser scanning is an imaging technique where a laser line moves on the target or the target moves under a laser line. At the same time, the camera images the different shapes of the laser line as the line shape changes according to the target (Figure 7.12) (Batchelor 2012).

When the camera imaging and the move of the target or the camera are synchronized, the imaged laser lines can be combined with the analyzing software. The real 3D model of the object can be presented and analyzed with these combined lines. There are some limitations of this technique. For example, the 3D shapes can be detected only from the top plane of the object. In other words, if there are any shape changes on the bottom plane, they are not recognized. Sometimes, the need to move the camera or the target may cause problems with this technique. In addition, it may be difficult to illuminate deep grooves and image at the same time.

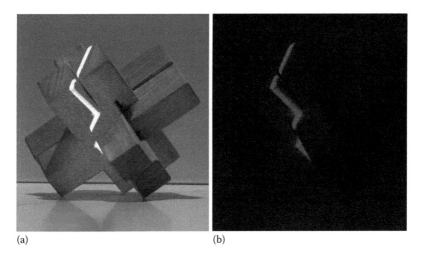

(a)   (b)

FIGURE 7.12 3D imaging with laser scanning. (a) The laser line on the surface of the target object. (b) The shape of the laser line as the camera sees it.

7.2.3.2 Stereo Imaging

Stereo imaging is a more complex 3D imaging technique. It uses two or more cameras to image the same target from somewhat different viewpoints. A 3D image is created by combining the images taken from different angles. Combining images is a multiphase and complicated process. First, the correspondence of the matching pairs found in different images must be confirmed. Second, a disparity map can be formed with the correspondent points. Next, if stereo geometry is known, the disparity map can be converted into a 3D model (Marshall 1994).

7.2.3.3 Kinect-Based 3D Imaging

Kinect-based 3D imaging technology was developed for the Microsoft Xbox game console. It is used in game plays to monitor the players' movements in three dimensions. Because of the low price of the Kinect camera, it has become very popular in different kinds of consumer applications. The imaging technique of Kinect is based on two separate cameras: the traditional type of camera that takes images in visible light wavelength and the other camera that takes images in near-IR wavelengths. The 3D imaging is based on imaging the changes in a near-IR dot pattern that is reflected to the target surface. The target point distances can be calculated based on this information. Microsoft has released a software developer kit (SDK) for Kinect. It is possible to code your own applications with it. The Kinect camera can also be purchased separately, without a game console (Figure 7.13) (Kortelainen et al. 2013).

7.2.3.4 Fringe Imaging

In 3D imaging called fringe imaging, the target is illuminated with different lighting patterns. The shape of the target items can be detected on the basis of the changes in the lighting pattern. An example of a lighting pattern is presented in Figure 7.14.

FIGURE 7.13 Kinect camera for application development (back) and for game console (front).

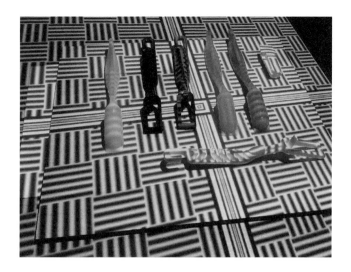

FIGURE 7.14   Fringe structural lighting pattern on the target surface.

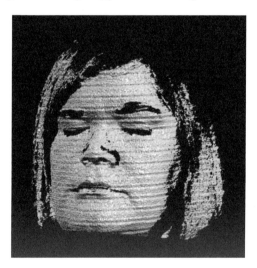

FIGURE 7.15   3D model of a human face by fringe imaging.

Figure 7.15, on the other hand, presents a 3D model of a human face made with a fringe-based imaging system (Batchelor 2012, Leino et al. 2015).

7.2.3.5   Time of Flight

Time-of-flight technology is based on measuring the time of flight of a light signal traveling between the camera and the target subject. It measures the time it takes for the light to travel from the camera to the target and back to the camera for each point of the image. In other words, it measures the light phase shift between the transmitted and received lights. A new-generation Kinect camera for the Xbox One game console is based on time-of-flight technology (Knies 2013, Kortelainen et al. 2013).

3D imaging technologies were mainly developed for industrial use, but they can also be applied to games, as shown by Kinect technology used in the Xbox game console. Commercial applications of these technologies are reasonably inexpensive and therefore more applicable in health and well-being sectors, too (Leino et al. 2015).

7.2.4 Nonvisible Machine Vision Technologies

Machine vision technologies described in earlier chapters usually refer to machine vision in visible light wavelengths, that is, machine vision solutions, which see about the same things as humans see. There are many imaging technologies for nonvisible wavelengths, and they can be integrated with the machine vision analyzing software to work automatically. This chapter concentrates on describing various special, non visible wavelength machine vision technologies such as IR imaging and spectral imaging.

7.2.4.1 Infrared Imaging

Infrared imaging is one of the best-known special machine vision system in non-visible wavelengths. Most of the research and applications still concentrate not only on long-wave infrared imaging, that is, thermal imaging, but also on near-IR imaging; near-IR imaging is already well known because of its special features. Thermal imaging has traditionally been used in finding hot or cold spots in processes, items, or environments. Lately, thermal imaging research has focused especially on auto-mated quality control, on heating process optimizations, on search and examination of welding or casting faults (Figure 7.16), and on identifying contact resistance in

FIGURE 7.16 Faults in welding detected through thermal imaging.

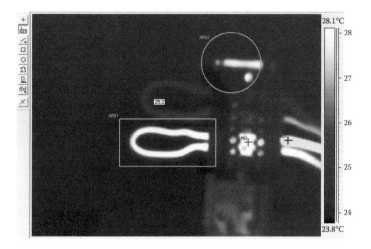

FIGURE 7.17 Identifying contact resistance in electrical connections with thermal imaging.

FIGURE 7.18 Near-infrared imaging used in moisture detection and in seeing through a kerosene-based liquid.

electrical connections (Figure 7.17). Near-IR imaging has been used mainly in moisture detection and in seeing through some materials invisible to the naked eye, for example, kerosene-based liquids, as shown in Figure 7.18 (Leino et al. 2015).

7.2.4.2 Near-Infrared Spectral Imaging

Near-IR spectral (NIRS) imaging is another type of very effective special machine vision technology. It is an innovative combination of spectroscopy, machine vision, and signal processing. Spectral imaging reveals the spectrum of the material at a certain point of an object from every pixel of the image. The spectrum also reveals the real color of the target object in visible light wavelengths. In the near-IR wavelengths, the spectrum is comparable to a fingerprint, identifying the concerned surface material

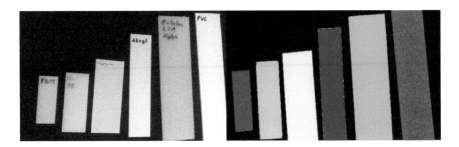

FIGURE 7.19   Plastics identified with near-infrared spectral imaging.

exactly. The spectrum identifies and visualizes each distinguished material that the target contains. Therefore, it is useful in identification and sorting of substances. An example of plastic identification is presented in Figure 7.19 (Leino et al. 2015).

7.3   MACHINE VISION FOR HEALTH AND SOCIAL CARE

As mentioned in Section 7.1, machine vision has traditionally been used in industrial applications. Its application to other fields has just started. The use of machine vision will increase with the proliferation of different kinds of service robots and automated assisting applications. The same development can be seen in smart health and social care. This chapter introduces a couple of machine vision applications designed for health and social care.

7.3.1   Separating Autism Spectrum Disorders with 3D Imaging

Autism is a neurodevelopmental disorder and consists of a spectrum of closely related disorders. Autism is diagnosed in patients who have problems with social interactions or in verbal or non verbal communication. They may also have restricted and repetitive behavior. It is very important to detect the symptoms of autism as early as possible to provide the most effective treatment for the best outcomes.

Obafemi-Ajayi et al. (2014) report how 3D facial imaging was used to detect the specific facial traits of autistic children. This 3D imaging system and the software designed for this purpose are considered good methods in defining the facial expressions that are common to children with autism. The imaging system is used to collect 3D coordinates of 19 facial landmarks. The distances between all possible pairs of these landmarks are then measured. As a result, the researchers captured 171 facial distance features. Complex software algorithms were used to categorize these viable biomarkers of autism spectrum disorders in subgroups.

7.3.2   Kinect-Controlled Physiotherapy

In physiotherapy, there are many treatments that are usually performed by physiotherapists manually. These treatments are often repeated on a daily basis for a prescribed period. Clinica Ordonez and Robotics Special Applications developed a robotic physiotherapy machine that can treat back problems in cervical, dorsal, and lumbar regions. The patients are automatically identified by their personal radio-frequency identification cards and the treatment is started. The Kinect-based imaging system is used to scan the patient's body shapes on the treatment area. The treatment can be carried out with the robot according to the physician's orders, with the help of the 3D model of the body (Riestra 2015).

7.3.3 AUTOMATED VENIPUNCTURE

Venipuncture aka venepuncture or venopuncture in medicine refers to obtaining an intravenous access for blood testing. Veebot (2014) is an automated venipuncture machine that combines machine vision and robotics. Veebot includes a table with an inflatable cuff, which fastens the patient's arm to the table and restricts blood flow, so that it is easier to detect the vein. Next, an infrared light is used to light the inner elbow, and the camera images it. The most usable vein is found from these images with the help of analyzing software. After that, an ultrasound is used to make sure that the vein is large enough and that there is adequate blood flow through the vein. Based on this analysis, the robot takes care of the needle aligning and of sticking the needle in. All these procedures take 1 minute, and a human is needed only for attaching the right test tube (Perry 2013).

7.4 MACHINE VISION CREATING FUTURE POSSIBILITIES FOR SMART HEALTH AND SOCIAL CARE

The earlier examples of machine vision in smart health and social care prove that there really is a clear need for new applications in this field. This subchapter compiles examples of the future possibilities of machine vision in smart health and social care. It introduces different applications under construction and considers the possibilities of different machine vision technologies for smart health and social care.

7.4.1 KINECT-BASED IMAGING IN ENHANCING WELL-BEING

The low-cost Kinect camera for Xbox 360 was designed for game playing, but it can also be used as a personal trainer. Microsoft released the SDK for application developers, offering a chance to create personalized application software. Kinect can be taught to give instructions based on observing the movements while exercising. Useful and personalized instructions on exercising correctly motivate to increase physical exercise.

An interesting example of 3D imaging applications designed for enhancing well-being is a Kinect-based human user interface. This application can be used to save personal movements in particular functions such as switching on or off electrical devices at home. For example, by lifting the right arm, the lights switch on, whereas by lifting the left arm, the light goes off again (Figure 7.20). This human user interface is usable as assistive technology or as a motivator in physical exercise. It is very easy to teach new movements to the Kinect application. In addition, it is easy to install the additional control electronics with interfaces to control intended systems (Leino et al. 2015).

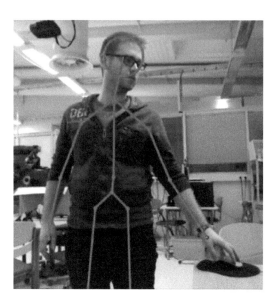

FIGURE 7.20   The Kinect application recognizes the user lifting his left arm and turns the lights off.

7.4.2   INFRARED IMAGING IN WELFARE TECHNOLOGY APPLICATIONS

The special infrared imaging technologies have also been used in welfare technology applications. Near-IR imaging is applied and studied, for example, in perspiration detection and thermal imaging for detecting muscle temperature during and after exercising.

Figure 7.21 presents the findings of a very interesting study, where different cold therapy methods were tested and their effects on muscles were examined by thermal imaging. Three students used different cooling methods, which each applied on the calf of their left leg. The first student used cold gel (a), the second used cold spray (b), and the third used a cold pack (c). The right-side calves were imaged as the control reference.

The effects of various cooling methods are clearly visible through thermal imaging (Figure 7.21). The legs were imaged 15 min after the cooling method was applied. The findings indicate that cold gel and cold spray had no cooling effects at all. Only the temperature in the calf handled with a cold pack had cooled down measurably, even if in

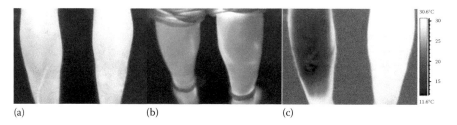

FIGURE 7.21   Three different pairs of calves, where each left calf was cooled down with a different muscle-cooling method. (a) Cold gel; (b) cold spray; and (c) cold pack.

all cases the persons felt cooling sensations in their calves. The sensed effects can be explained by the fact that the gel and spray irritated nerve sensors on the skin and sent messages to the brain that the muscle is cooling down. This study confirmed only the measurable temperature changes in the muscle tissue. However, it could not be detected whether the feeling of cooling down caused any changes in the muscle (Leino et al. 2015).

7.4.3 NEAR-INFRARED SPECTRAL IMAGING IN DIAGNOSTICS

The usability of NIRS imaging in diagnostics has been researched in near-IR camera applications, for the use in the interface of welfare technology and electronics project at Satakunta University of Applied Sciences. The goal of the project was to find out if and how NIRS imaging could be used in inflammation detection on human skin or just under it. The secondary research question concerned about the possibilities of the technology to work better and more predictably than human eyes. In this project, the blood flow changes were detected, for example, in inflamed tissues, or skin grafts were researched through NIRS imaging. Figure 7.22 presents how severely inflamed tissues are shown on a person's left hand. Similar inflamed parts in terms of blood flow change are red in the spectral image of the right hand, where only some bruising could be detected by naked eyes (Huhtanen 2007, Leino et al. 2015).

Figure 7.23 presents a skin graft. The green color in an NIRS image represents deteriorated blood flow and blue color represents normal blood flow. Analyzed by NIRS image, these qualifications indicate that one part of the skin graft has healed better than the other (Leino et al. 2015).

FIGURE 7.22 Septic hand (left hand) and a beginning bruise in the right hand in near-infrared spectral imaging.

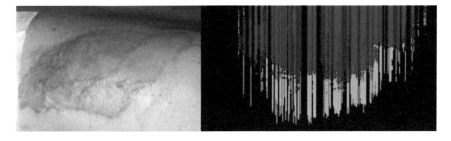

FIGURE 7.23 Skin graft in nature and in near-infrared spectral image.

7.5 CONCLUSION

Machine vision applications and their research have earlier been intended for industrial use. Today, there are several examples of how machine vision technologies known from industrial applications are also used in enhancing well-being. The advantages of machine vision compared with traditional methods have already been recognized in diagnostics or physical therapy, in particular, noninvasiveness and the possibility to image large areas simultaneously. Traditional machine vision technologies, particularly different kinds of special machine vision technologies, not only enable monitoring of health and well-being, for example, physical and physiological changes in a human body, but they also make it possible to recognize features and phenomena that humans cannot see. The well-known industrial base of machine vision creates opportunities to apply these technologies to new, innovative applications in completely different areas.

REFERENCES

Automated Vision Inc. 2006. Fundamentals of machine vision. http://www.autovis.com/web/courses/Fundamentals1/Presentation_Files/index.html (accessed on July 08, 2016).

Batchelor, B.G. (ed.) 2012. Illumination sources. In *Machine Vision Handbook*. New York: Springer, 284–316.

Huhtanen, J. 2007. *Methods of Analysis of Spectral Images in Patient Diagnostics*. Master of Science Thesis. Tampere University of Technology.

Knies, R. 2013. Collaboration, expertise produce enhanced sensing in Xbox One. The official Microsoft blog. http://blogs.microsoft.com/blog/2013/10/02/collaboration-expertise-produce-enhanced-sensing-in-xbox-one/ (accessed on July 08, 2016).

Kortelainen, J., Leino, M., Valo, P. 2013. The advanced 3D imaging techniques (3D-kuvauksen kehittyneettekniikat [in Finnish]). *Automaatioväylä*, 29(6): 12–14.

Leino, M., Valo, P., Kortelainen, J., Laine, K. 2015. Tackling the challenges of well-being with machine vision. In: Sirkka, A. (ed.) *Arts, Games and Sensors Harnessed to Enhance Well-being*. Satakunta University of Applied Sciences, Series B, Reports 3/2015. http://www.theseus.fi/bitstream/handle/10024/89158/2015_D_3_SAMK_Art_Games_and_Sensors_low.pdf?sequence=2 (accessed on July 08, 2016).

Marshall, D. 1994. Introduction to stereo imaging—Theory[Online]. Cardiff School of Computer Science & Informatics. Cited January 4, 2014. http://www.cs.cf.ac.uk/Dave/Vision_lecture/node11.html.

Obafemi-Ajayi, T., Miles, J.H., Takahashi, T. N., Qi, W., Aldridge, K., Zhang, M., Xin, S-Q., He, Y., Duan, Y. 2014. Facial structure analysis separates autism spectrum disorders into meaningful clinical subgroups. *Journal of Autism and Developmental Disorders*, 45(5): 1302–1317.

Perry, T.S. 2013. Profile: Veebot—Making a robot that can draw blood faster and more safely than a human can. *IEEE Spectrum*. [Online], http://spectrum.ieee.org/robotics/medical-robots/profile-veebot (accessed on July 08, 2016).

Riestra, I. 2015. Vision helps robot perform physiotherapy. *Vision Systems Design*. http://www.vision-systems.com/articles/print/volume-20/issue-1/features/vision-helps-robot-perform-physiotherapy.html?cmpid=EnlVSDApril62015 (accessed on July 08, 2016).

Veebot, LLC. 2014. Automated venipuncture solutions[Online]. http://www.veebot.com/solutions.html (accessed on July 08, 2016).

8 The Possibilities and Challenges of RFID-Based Passive Wireless Components in Healthcare Applications

Johanna Virkki, Sari Merilampi, and Serafin Arroyo

CONTENTS

8.1 INTRODUCTION

In the healthcare and well-being industry, technology applications answering the future challenges are gaining more and more interest. For example, the lack of personnel in nursing homes is a significant problem, and the situation is not getting any easier. The current trend is also to help the elderly to live at home instead of caring homes. In addition, as the importance of home nursing is expected to grow in general, the freedom of movement achieved by technology is especially significant. Rather than replacing human contact, technology should be assistive and take care of certain monitoring tasks, thus saving the time of the professional staff for more important tasks. Technology should also provide security and confidence to live in full capacity. In this chapter, we introduce the possibilities and challenges of radio frequency identification (RFID)-based applications in healthcare and well-being contexts.

The RFID technology was designed for automatic identification or tracking of objects and people. The widespread use of RFID in other industries has resulted in attempts to introduce applications in the healthcare environment also, such as

RFID-based automatic patient identification systems. Patient identification and the matching of a patient to an intended treatment are the activities that are performed routinely in all care settings. Risks to patient safety occur when there is a mismatch between a certain patient and the components of care, whether these components are diagnostic, therapeutic, or supportive. The failure to identify patients correctly may result in wrong site procedures, medication errors, transfusion errors, and diagnostic testing errors. The RFID technology has the potential to remove these kinds of problems by providing an improved identification system. It also has a lot of potential in tracking and identifying important equipment. The technology can also be further developed and applied to other purposes, such as wearable wireless sensing, which also has versatile possibilities in healthcare and well-being contexts.

Several RFID systems have been implemented and are available in the market. We will focus on passive RFID technology, in which the used simple tags do not have any power source on them and thus have almost unlimited lifetime and low price. Use of passive RFID technology requires wirelessly readable electronic tags, readers, and background systems:

- *Tags*: These are composed of an antenna for communication and an integrated circuit (IC) that stores the information of the item/person to which the tag is attached.
- *RFID reader*: It is responsible for powering the tags and communicating with the tags. The communication and coupling between the reader and the tags in passive systems are based on radio frequency electromagnetic fields (high-frequency [HF] systems) and waves (ultrahigh-frequency [UHF] systems).
- *Background system*: The readers are connected to a network in which the data are collected, processed, stored, used, and shared.

During this chapter, we will explore two different kinds of passive RFID technologies: HF RFID and UHF RFID. In case of HF RFID technology, we will focus on one technology that has progressed a lot during the last few years, that is, near field communication (NFC). We will introduce the use of NFC in patient identification and assets tracking, the two most common applications of RFID in the industry. We will then continue with UHF RFID technology, where we will discuss more novel applications of RFID, such as passive wireless sensors. Further information on RFID technology and a comprehensive introduction to today's systems are provided in (Want, 2006; Dobkin, 2008).

8.2 PASSIVE NFC TECHNOLOGY IN PATIENT IDENTIFICATION AND ASSETS TRACKING

Near field communication is designed to be a secure identification system and also a secure form of data exchange. An NFC device is capable of being both an NFC reader and an NFC tag. This unique feature allows NFC devices to communicate peer to peer, as well as provide card emulation capabilities. The NFC systems operate at

13.56 MHz frequency and have read ranges up to 4 cm. One of the main advantages of NFC is the availability of NFC reader in many smartphones and tablets. The major disadvantage is the short read range.

The NFC technology can be used for reliable patient identification systems in hospitals and caring homes. The system helps avoid errors, for example, if a healthcare professional forgets to mark in a log that the patient was visited or medicated. The use of NFC in patient identification leads to the following process: When a patient arrives to a hospital, an NFC tag is provided to the patient, for example, by attaching a wristband. Doctors and other staff members will be equipped with NFC-enabled smart phones or NFC readers, and they will use a software app that interprets the tag data regarding patients and the medications they receive. All information is stored in a server, and staff members can easily access it through the Internet. As RFID tags require no line of sight, they can be read, for example, through bed sheets, without disturbing the sleeping patient, a major advantage over using a barcode wristband.

In healthcare environments, NFC can also be used for inventory purposes. The solution would consist of a management system with a web interface that uses an NFC reader to read NFC tags placed on the different assets. This system is fully scalable and allows to uniquely identify all types of assets. By using an NFC solution, the hospital staff can create and organize their inventory and manage the asset issues efficiently. In addition, the maintenance processes of the hospital assets are simplified, resulting in reduced costs.

8.3 PASSIVE UHF RFID TECHNOLOGY IN WIRELESS SENSING

As NFC systems, passive UHF RFID systems allow automatic identification of objects remotely. The use of propagating electromagnetic waves in the UHF frequency range (center frequencies vary globally but fall within the range from 860 to 960 MHz) enables rapid interrogation of a large number of tags through various materials, and the tag can be read from distances of over 10 m, which in practice is the main difference when compared with NFC. Similarly to NFC, the first step in utilizing passive UHF RFID tags in healthcare and well-being areas is not only to track assets but also to monitor, track, and identify people, for example, patients who carry tags in their wristbands. With read ranges of several meters, UHF RFID tags enable monitoring of people without their conscious effort.

Instead of just tracking people and equipment, passive UHF RFID tags offer more possibilities. The UHF RFID tags are promising candidates to work as energy-autonomous wireless sensors that exhibit low complexity and cost. There are two types of passive UHF RFID sensor tags:

1. The RFID tag in which a traditional sensor is attached as a part of the tag or integrated into the tag IC.
2. The RFID tag in which the sensing ability is integrated into the tag structure.

In the former type, the RFID tag is typically used in power supply and data transfer, and in the latter, the tag performs the sensor function (Merilampi et al., 2012). Novel RFID ICs have input/output (I/O) ports for external sensors, and there are also ICs available with integrated sensors, such as a temperature sensor, a pressure sensor, and a moisture sensor. The selection of integrated sensors is still very limited, but the use of external sensors substantially widens the sensing possibilities. However, external sensors may require battery-assisted tags to be used for providing enough supply power.

In the latter type, it is possible to establish maintenance-free sensors without external sensors or onboard electronics by using a passive UHF RFID tag antenna as the sensing element. Antenna-based sensing provides integration of sensing capabilities in passive RFID tags, with a minimal increase in the overall complexity and power consumption of the tag. These types of simple tags can also be integrated into clothes for wireless wearable sensors.

The UHF RFID readers can be embedded, for example, in the hospital bed, or they can be hand-held devices. The data can be queried at predefined intervals and transmitted to a background system for automatic processing or only on demand for the benefit of the user.

In addition to UHF RFID, lower-frequency systems could also be used for sensing purposes. If short read ranges are sufficient, the previously introduced NFC technology offers an interesting choice. The NFC ICs allow external sensors and electronics to be used and thus also offer interesting platform for wearable sensing. Since NFC technology has the benefit of being accessible with a mobile phone, all the sensor data can be directly analyzed and sent further. It is also worth mentioning that although UHF RFID readers do not yet exist in mobile phones, promising development is going on in reducing the size of the hand-held reader and connecting the reader to a mobile phone. Already very small readers are available, and they can be connected to a mobile phone wirelessly (e.g., via Bluetooth). Next, we will concentrate on long-range systems, and NFC is no further discussed.

8.4 POSSIBILITIES AND CHALLENGES OF PASSIVE UHF RFID SENSOR TAGS

Although passive UHF RFID technology provides many possibilities, the use of this technology for sensing is still in early stage. There are still challenges to overcome, such as optimization of sensor designs, development of efficient fabrication methods and optimized materials, and ensuring the reliability in actual fields use.

In addition to the traditional manufacturing methods, such as etching, versatile methods have been studied for tag antenna fabrication: Embroidery with conductive yarns is a simple fabrication method with a high potential in antenna fabrication (Kaufmann et al., 2013; Tsolis et al., 2014) and can be particularly useful method when embedding interconnections into textiles. The use of electrotextiles based on conductive threads and metal-coated fabrics has been studied, and they have been found suitable for antenna fabrication (Virkki et al., 2015). In addition, different additive methods have been used for tag manufacturing: Brush painting of conductive inks on fabric is a simple but versatile fabrication method of antennas (Virkki et al., 2014). Screen printing (Merilampi et al., 2011; Virkki et al., 2015) and inkjet printing (Whittow et al., 2014) of conductive inks have also been successfully used for textile antenna fabrication, to name some examples. Three-dimensional direct write depositing of conductive inks has the potential to enable next-generation manufacturing of antennas and interconnections, and some results already exist to demonstrate the potential of this method (Björninen et al., 2015).

The RFID tag's response can be manipulated according to the prevailing circumstances such as mechanical changes in the tag or presence of certain materials, for example, water. The changes in the antenna or antenna substrate affect the impedance matching of the tag antenna and the IC, as well as the gain of the tag antenna. Through these, they also affect wirelessly measureable parameters, such as the transmitted threshold power, the power on tag, and the backscattered power of the tag. The transmitted threshold power is the minimum sufficient power required from the RFID reader to activate the IC. The backscattered signal power is the time-average power detected from tag response at the receiver. The power on tag is the measure of power available for the tag when the tag "awakes." It can be interpreted as the power that would be absorbed by the RFID chip if it was connected to a perfectly matched 0 dBi tag antenna.

Several prototypes of RFID sensor tags have been presented, and interesting applications have been suggested in versatile fields. As mentioned, by using optimized antenna designs, the tag antenna itself may be used as a sensor, for instance, a gas sensor tag (Occhiuzzi et al., 2011) based on a carbon-nanotube-loaded antenna and a moisture sensor tag (Virtanen et al., 2011), where the distributed capacitance in the antenna varies with the amount of moisture absorbed by the substrate. High-permittivity polymer ceramic was used as a temperature-sensitive substrate for the sensor antenna in (Qiao et al., 2013). In addition, a threshold temperature sensor tag, where paraffin wax was used as a substrate that experiences an irreversible state change after a threshold temperature, has been presented (Babar et al., 2012). A crack propagation sensor tag, where the antenna resonance frequency was correlated with the length of a crack, has been investigated (Yi et al., 2014). One of the first wearable sensor components, a strain sensor tag, based on the backscatter strength of a screen-printed stretchable antenna, has been introduced (Merilampi et al., 2011). In addition, an RFID strain sensor tag, based on an electrotextile antenna with a stretchable section that modifies the antenna properties as a function of the strain, has been presented (Long et al., 2015). Next, we will introduce the possibilities of two interesting applications of RFID-based passive wireless sensors: wearable moisture-sensor tags and wearable strain-sensor tags.

8.4.1 MOISTURE SENSOR

In case of RFID-based moisture sensor, the wireless response of the tag as a function of increased humidity is measured. The increased moisture will change the permittivity of the fabric substrate (and thus change the impedance of the antenna, which will create the mismatch between the tag antenna and the IC) and also increase the ohmic losses in the fabric substrate, degrading the overall tag performance. An example of printed UHF RFID tags' "power on tag" before and after high-humidity conditions is presented in Figure 8.1.

Wearable humidity sensors have versatile applications in the healthcare field. For example, in wound healing, real-time moisture measurements will provide an indication when to replace the dressing. Moreover, the basic problem of keeping

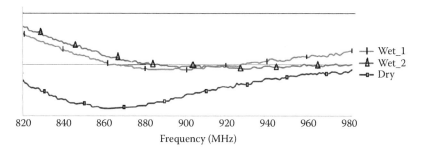

FIGURE 8.1   The change in the "power on tag" of printed passive UHF RFID tags on paper substrate after high-humidity conditions.

checking whether a diaper needs changing can be avoided by wireless moisture sensors. Further, as many physiological and metabolic functions are affected by an individual's hydration status, sweat rate can be used to estimate body hydration and the amount of fluid needed. A person's sweat rate also relates to exercise intensity, environmental conditions, and acclimatization state. Hydration, performance, and physiological factors are all interrelated. Therefore, by monitoring sweat loss and composition, in addition to heart rate and breathing, it is possible to get a more complete picture of the body's physiological state. In addition, certain diseases and seizures cause sweating (Candas et al., 1986; Maughan et al., 2007), which could be immediately noticed by wireless humidity sensors.

8.4.2 Strain Sensor

The effect of mechanical straining on RFID tag functioning can be observed based on the signal received from the tag. The response of the tag as a function of strain is dependent on the tag geometry, the tag antenna, and substrate materials (Long et al., 2015). An example of passive UHF RFID strain-sensor response is presented in Figure 8.2.

Wearable strain sensors will allow remote monitoring of physiotherapy exercises, posture, and flexibility during normal everyday tasks. Moreover, they can be used to provide real-time feedback to neurological patients undergoing motor rehabilitation. After further development, strain sensor tags could be used in various health-care applications in, for example, monitoring the movement of healing limbs (to warn about too large a movement) or monitoring chest movements (breathing). Other applications could be found in rehabilitation and in exercising to prevent diseases. The sensor could be used for counting the movements during a workout and in exercise games. In movement-monitoring applications, the sensor could be embedded in clothes in places that experience strain (e.g., elbows, knees, ankles, and wrapped around the chest), or it could be used in a separate belt. In exercise or game applications, the sensor could be integrated into exercise equipment or clothes, or it could be used as the game controller. The sensor could also be used in measurements of

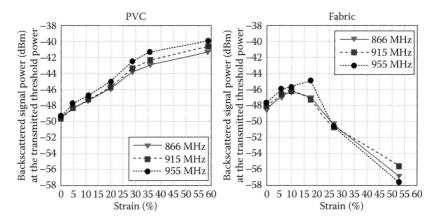

FIGURE 8.2 The backscattered signal power of a strain sensor tag on a polyvinyl chloride substrate as a function of strain at 866, 915, and 955 MHz. (From Merilampi, S. et al., *Sens. Rev.*, 31, 32–40, 2011. With permission.)

force, since the strain is proportional to the force (Lorussi et al., 2005; Giorgino et al., 2006; Merilampi et al., 2012; Tormene et al., 2012).

Despite the definite advantages and high potential, some challenges related to the effective realization and reliability of these passive wireless components must be solved before large-scale deployment of wearable passive RFID sensors.

The design of effective wearable tags for UHF RFID applications involving people is still a major challenge, owing to the strong interaction of the antenna with the human body, which causes impedance detuning and efficiency degradation (Occhiuzzi et al., 2010). Another design challenge considering these self-sensing tags is the designing of the "sensor component" in the tag geometry. One major demand imposed to body-centric passive sensor systems is to operate in unobtrusive way, that is, that the user is not aware of the presence of these sensors. Furthermore, in real conditions, on the moving human body, antennas are prone to bending and twisting, which is another constraint that needs to be taken into account in antenna design and also when positioning them on the body (Hall and Hao, 2006).

In addition, the reliability of these sensors in changing (not to mention challenging) use conditions (Lilja et al., 2012) is another major issue to be addressed. From reliability point of view, the most fragile part that typically leads to malfunctions is (component) joint, which may break during use. Thus, the minimum number of component joints is preferred. During actual use, wearable RFID tags have to endure many kinds on environmental stresses, such as changing and harsh weather conditions (Paul et al., 2014) and versatile mechanical stress, including stretching (Virkki et al., 2015) and bending (Kaufmann et al., 2013; Paul et al., 2014). Especially machine washing (Kellomäki et al., 2012; Yan et al., 2015) has been found to be a major reliability challenge for the electrical performance of electrotextile components. The challenge remains in identifying suitable low-cost conformal coating materials that are compatible with textile materials and do not disturb the performance of the sensor tags.

8.5 CONCLUSIONS

The RFID technology may be utilized in the healthcare and well-being industry. Most common applications are identification and tracking systems of assets and people. However, other applications are also emerging. The RFID-based wearable sensors have enormous potential and versatile future applications. The possibilities of these wireless components have been presented in this chapter, together with the recent scientific results from the field. However, some challenges still hinder the effective application of such technology to nursing homes and hospitals, and these have also been highlighted.

REFERENCES

Babar, A.A., Manzari, S., Sydänheimo, L., Elsherbeni, A.Z., and Ukkonen, L. 2012. Passive UHF RFID tag for heat sensing applications, *IEEE Trans Antennas Propag*, 60(9):4056–4064.

Björninen, T., Virkki, J., Sydanheimo, L., and Ukkonen, L. 2015. Possibilities of 3D direct write dispensing for textile UHF RFID tag manufacturing. In *IEEE International Symposium on Antennas and Propagation & USNC/URSI National Radio Science Meeting*, IEEE, Vancouver, Canada, July 19–24, pp. 1316–1317.

Candas, V., Libert, J.P., Brandenberger, G., Sagot, J.C., Amoros, C., and Kahn, J.M. 1986. Hydration during exercise. Effects on thermal and cardiovascular adjustments, *Eur J Appl Physiol Occup Physiol*, 55(2):113–22.

Dobkin, D. 2008. *The RF in RFID: Passive UHF RFID in Practice*. Newnes-Elsevier, Amsterdam, the Netherlands.

Giorgino, T., Lorussi, F., De Rossi, D., and Quaglini, S. 2006. Posture classification via wearable strain sensors for neurological rehabilitation, *Conf Proc IEEE Eng Med Biol Soc*, 1:6273–6276.

Hall, P.S. and Hao, Y. 2006. *Antennas and Propagation for Body-Centric Wireless Communications*. Artech House, Boston, MA.

Kaufmann, T., Fumeaux, I.M., and Fumeaux, C. 2013. Comparison of fabric and embroidered dipole antennas. In *European Conference on Antennas and Propagation*, Gothenburg, Sweden, April 8–12, pp. 325–3255.

Kellomäki, T., Virkki, J., Merilampi, S., and Ukkonen, L. 2012. Towards washable wearable antennas: A comparison of coating materials for screen-printed textile-based UHF RFID tags, *Int J Antennas Propag*, 11 p. Article ID 476570. doi:10.1155/2012/476570.

Lilja, J., Kaija, T., and De Maagt, P. 2012. Design and manufacturing of robust textile antennas for harsh environments, *IEEE Trans Antennas Propag*, 60:4130–4140.

Long, F., Zhang, X., Björninen, T., et al. 2015. Implementation and wireless readout of passive UHF RFID strain sensor tags based on electro-textile antennas. In *European Conference on Antennas and Propagation*, Lisbon, Portugal, April 13–17, pp. 1–5.

Lorussi, F., Scilingo, E.P., Tesconi, M., Tognetti, A., and De Rossi, D. 2005. Strain sensing fabric for hand posture and gesture monitoring, *IEEE Trans Inf Technol Biomed*, 9:372–381.

Maughan, R.J., Shirrefes, S.M., and Leiper, J.B. 2007. Errors in the estimation of hydration status from changes in body mass, *J Sports Sci*, 25(7):797–804.

Merilampi, S., Björninen, T., Sydänheimo, L., and Ukkonen, L. 2012. Passive UHF RFID strain sensor tag for detecting limb movement, *Int J Smart Sens Intell Syst*, 5(2):315–328.

Merilampi, S., Björninen, T., Ukkonen, L., Ruuskanen, P., and Sydänheimo, L. 2011. Embedded wireless strain sensor based on Printed RFID tag, *Sens Rev*, 31(1):32–40.

Occhiuzzi, C., Cippitelli, S., and Marrocco, G. 2010. Modeling, design and experimentation of wearable RFID sensor tag, *IEEE Trans Antennas Propag*, 58(8):2490–2498.

Occhiuzzi, C., Rida, A., Marrocco, G., and Tentzeris, M.M. 2011. RFID passive gas sensor integrating carbon nanotubes, *IEEE Trans Microw Theory Tech*, 59(10):2674–2684.

Paul, G., Torah, R., Yang, K., Beeby, S., and Tudor, J. 2014. An investigation into the durability of screen-printed conductive tracks on textiles, *Meas Sci Technol*, 25(2):11.

Qiao, Q., Zhang, L., Yang, F., Yue, Z., and Elsherbeni, A.Z. 2013. UHF RFID temperature sensor tag using novel HDPE-BST composite material. In *IEEE International Symposium on Antennas and Propagation & USNC/URSI National Radio Science Meeting*, IEEE, Orlando, FL, July 7–13, pp. 2313–2314.

Tormene, P., Bartolo, M., Nunzio, A., et al. 2012. Estimation of human trunk movements by wearable strain sensors and improvement of sensor's placement on intelligent biomedical clothes, *Biomed Eng Online*, 11:95. doi:10.1186/1475-925X-11-95.

Tsolis, A., Whittow, W.G., Alexandridis, A., and J. Vardaxoglou, J. 2014. Embroidery and related manufacturing techniques for wearable antennas: Challenges and opportunities, *Electronics*, 3(2):314–338.

Virkki, J., Björninen, T., Merilampi, S., Sydänheimo, L., and Ukkonen, L. 2015. The effects of recurrent stretching on the performance of electro-textile and screen-printed ultra-high-frequency radio-frequency identification tags, *Text Res J*, 85(3):294–301.

Virkki, J., Björninen, T., Sydänheimo, L., and Ukkonen, L. 2014. Brush-painted silver nanoparticle UHF RFID tags on fabric substrates. In *Progress in Electromagnetics Research Symposium*, European Association on Antennas and Propagatio, Guangzhou, China, August 25–28, pp. 2106–2110.

Virtanen, J., Ukkonen, L., Björninen, T., Elsherbeni, A.Z., and Sydänheimo, L. 2011. Inkjet-printed humidty sensor for passive UHF RFID systems, *IEEE Trans Instrum Meas*, 60(8):2768–2777.

Want, A. 2006. An introduction to RFID technology, *IEEE Pervas Comput*, 5(1):25–33.

Whittow, W.G., Chauraya, A., Vardaxoglou, J.C., et al. 2014. Inkjet-printed microstrip patch antennas realized on textile for wearable applications, *IEEE Antennas Propag Lett*, 13:71–74.

Yan, F., Chan, Y., Ming, Y., et al. 2015. Experimental study on the washing durability of electro-textile UHF RFID tags, *IEEE Antennas Wireless Propag Lett*, 14:466–469.

Yi, X., Cho, C., Wang, Y., Cook, B., Tentzeris, M.M., and Leon, R.T. 2014. Crack propagation measurement using a battery-free slotted patch antenna sensor. In *European Workshop on Structural Health Monitoring*, July 8–11, Nantes, France, 8 p.

9 Acoustic-Based Technologies for Ambient Assisted Living

Maximo Cobos, Juan J. Perez-Solano, and Lars T. Berger

CONTENTS

9.1   INTRODUCTION

Over the past years, the world's population has continued its transition from a state of high birth and death rates to the one characterized by low birth and death rates. A consequence of this transition is currently the aging population, which is putting severe stress on modern social and medical welfare systems [1]. One way of tackling the problem is to prolong the time for which persons live in their own homes independently by using assistive technologies. To this end, information and communication technologies assisting the elderly are of high social and economic relevance. Examples of assistive technologies are reminder systems, medical assistance and tele-healthcare systems, personal emergency response systems, social robotics, and accessible human-machine interfaces [2]. However, when deploying such systems in a home, several considerations must be taken into account. On the one hand, the development of *ambient assisted living* (AAL) systems usually relies on application-dependent sensors, such as vital sensors, cameras, and microphones. On the other hand, since such systems are generally complex to use, the design of simple and intuitive interfaces is of great importance. It has been shown that natural and convenient ways to interact with technical systems are highly desired [3]. One way to develop an intuitive human-machine interaction is by using sound. The use of sound in AAL systems provides many advantages, since it has been shown that microphones can be easily integrated into existing living environments, they may serve multiple purposes (e.g., event detection, localization, and speech recognition), and users do not perceive them as obtrusive [4]. In fact, a monitoring device such as a video camera may be considered a severe break of privacy and, thus, is judged critically by the users. In contrast, acoustic monitoring can be considered a more accepted alternative. If the signal analysis is performed automatically by the system without storing acoustic data and without the possibility that other people may listen to the signals, this type of analysis would likely not be considered a break of privacy at all [5].

While a number of acoustic sensing and processing systems have been proposed over the last decades, these have typically relied on high-throughput computing platforms and/or expensive microphone arrays. Support for acoustic sensing is commonly found in many off-the-shelf solutions, although they vary greatly in terms of available facilities. A PC's hardware or laptops with multichannel audio interfaces offer great audio-related features, but their use tends to be more difficult in a distributed context, owing to cabling, volume, and other installation issues. The alternative is to use comparatively low-resource, distributed nodes with sensing devices and algorithms aimed at detecting, localizing, or characterizing acoustic events. The advantage of these systems is that the wireless, battery-powered nodes are less expensive and can be easily deployed in a wide range of environments. Moreover, as opposed to traditional microphone arrays that sample a sound field only locally, distributed acoustic sensing systems allow to use many more sensors to cover a large area of interest, as shown in Figure 9.1.

This is one of the main reasons why acoustic sensing by means of constrained devices has been receiving increasing attention in the last years. In fact, by taking advantage of appropriate sensors and signal processing, many innovative applications related to sound capture and analysis have emerged over a wide range of fields, including audio-based surveillance [6,7], intelligent auditory interfaces [8], smart teleconferencing [9], and habitat monitoring [10]. Among the variety of applications of acoustic sensor networks, an innovative one in the health domain is the promotion of smart homes, designed to support daily living tasks in the AAL contexts.

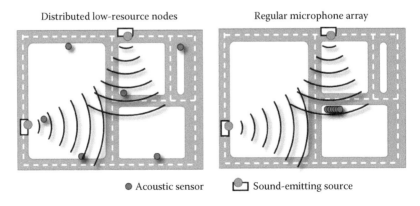

Distributed low-resource nodes   Regular microphone array

● Acoustic sensor   ▭ Sound-emitting source

FIGURE 9.1 Distributed acoustic network versus traditional microphone arrays.

These homes are envisioned to be equipped with a set of actuators, sensors, and computing platforms aimed at early detection of distress situations, remote monitoring, and the promotion of safety and well-being. Despite the great amount of sensing modalities available at a smart home, such as video cameras, RFID tags, and infrared sensors, audio signal processing technology offers some well-known advantages. First, since voice is the dominant human communication modality, interfaces aimed at disabled people or the aging population can be better adapted by using specifically designed audio features. Second, audio is a cheap and complimentary modality and does not require visual or physical interaction. Third, acoustic sensing devices can be especially useful in situations when other sensors fail to provide reliable environmental information (e.g., when a sound-emitting target is in the dark or is occluded). Figure 9.2 shows an example where a fall detector has the ability to detect the event even when the direct line-of-sight path is occluded. In addition, there are situations or events that can be more effectively detected using audio sensors when compared with image sensors, such as human shouting or crying and door knocking. Moreover, the combination of audio with other sensing modalities usually improves the overall system performance. As an example, the Sweet-Home project addressed some of the above issues by using audio-based technology to let users have full control over their home environment, as well as to ease the social inclusion of the elderly and frail population [11].

This chapter discusses current challenges, platform alternatives, and applications related to acoustic-based technologies in an assisted-living context. The chapter is intended to provide the reader with a broad view of potential uses of audio processing in the AAL domain, as well as to introduce the main computing platforms that

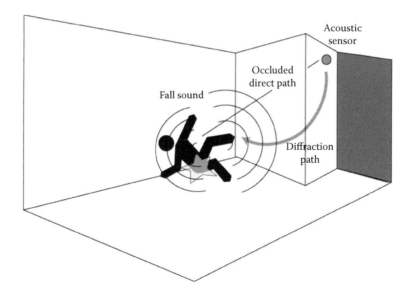

FIGURE 9.2 An acoustic sensor is able to obtain important information from an event of interest, even in situations where the sensor to object line of sight is obstructed.

may support these applications. The chapter begins with a description of general issues that must be considered in the design of *wireless acoustic sensor networks* (WASNs) from the software, hardware, and network levels. Then, the main computing platforms that have been used throughout the literature are presented, discussing the potential of current single-board computers for general-purpose acoustic sensing systems. Section 9.2 describes the most important audio processing tasks to be performed in AAL environments: acoustic source localization, acoustic activity detection (AAD), and audio event detection (AED). These audio processing tasks are considered key enabling technologies in acoustic-based AAL systems. While there are not widely accepted algorithms working under all type of situations, we comment on some common approaches found in the literature. Finally, some conclusions and further considerations are provided.

9.2 WIRELESS ACOUSTIC SENSOR NETWORKS

A WASN consists of a set of wireless microphone nodes, which are spatially distributed over the environment, usually in an ad hoc fashion. Connecting the nodes wirelessly, limitations related to the array size disappear, and the microphone nodes can physically cover a much wider area, increasing the amount of spatial information captured by the whole system.

The design of a WASN is very challenging, owing to several aspects. The obvious one is that real-time audio acquisition and processing adds significant data traffic in the network. As a result, there is a need for scalable solutions, both at the signal processing level and at the network-communication level. The core challenges that appear in the design of such a system are as follows [12]:

- *Unknown sensor geometry*: The locations of the nodes are not known a priori (usually owing to random deployment). Some acoustic processing tasks such as source localization and signal enhancement need to estimate the position of the different nodes in advance.
- *Distributed processing*: If there is no fusion center to collect the information from all the nodes, it is desired to distribute the computational load over the different nodes.
- *Bandwidth usage*: As bandwidth is a scarce resource, it is important to use it efficiently. If nodes share data only with their nearest neighbors, less transmission power is required.
- *Scalability*: In the design of algorithms for networks with many sensors, scalable features are desirable. In this context, adding extra nodes should not have a deep effect on the overall computational load or data traffic.
- *Sensor subset selection*: In large-scale networks, it might be sufficient to use only a subset of microphones to obtain the desired performance. The rest of nodes can then be put to sleep to save energy.
- *Input-output delay*: Real-time audio streaming is a challenging task. Delays are introduced both at the signal-processing level and at the network layer.
- *Synchronization*: Timing issues appear due to independent clocks at the nodes, resulting in an inevitable clock skew and offset. Note that clock

synchronization is very important not only at the data transmission layer, but also at data sampling layer. For many signal-processing algorithms, combining the outputs from sensors working at different sampling rates or under the effect of clock skew has a negative effect on the performance. Moreover, nonuniform networks, with devices from different manufacturers, can be much more problematic.

- *Routing and topology selection*: In data-intensive applications, the topology may be optimized in terms of transmission power, end-to-end delay, or quality of service, in general.

From a general point of view, research in acoustic sensing networks can be divided into two areas: hardware area (platforms and sensors) and software area (signal-processing algorithms, collaboration, and networking techniques). Both areas must be taken into account when designing a new application. For example, collaborative signal processing reduces not only the number of packets to be transported but also the probability of collision and interference in the shared media. To facilitate collaborative data processing, sensors are usually organized into clusters, as shown in Figure 9.3.

These clusters can be static (attributes such as the size of a cluster, the area it covers, and the members belonging to the cluster are static) or dynamic (formation of a cluster is triggered by a certain event, such as the detection of an approaching target with acoustic sounds). While the selected hardware platform must be easily deployable and offer the sufficient computational and communication resources, software flexibility is also required to support multiple concurrent applications, allow reconfiguration, and tune applications [13]. Moreover, an important aspect of acoustic processing in resource-constrained devices is that sound capture usually requires sampling signals at the Nyquist rate and transferring samples to higher-capability devices. This approach is bandwidth-intensive, because sensors must work at a high rate (10 kHz or above) and expend a lot of energy in data transfer. Fortunately, results from compressive sensing theory suggest that a small number of random samples

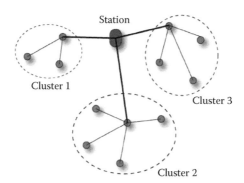

FIGURE 9.3 Collaborative networks share the total processing among clusters of sensors, reducing data flow and interference.

can embed the signal structures sufficiently well, provided that the signal is sparse in some domain [14].

On the hardware side, mobile platforms for acoustic sensing are constituted by three key elements: (1) low power processors; (2) wireless transceivers; and (3) acoustic sensors. Regardless the specific application, audio capturing and processing in these platforms can be considered a challenging matter, which involves a trade-off between the complexity of the audio processing task and the hardware resources. The audio monitoring process introduces some specific requirements on the hardware platforms, such as follows:

- Audio signals are normally sampled at relatively high rates, demanding large memories and high computational capabilities.
- Signal-processing tasks should be programmed carefully to deal with the audio sampling process and to optimize the system resources properly.
- Depending on the final application, either online or offline signal processing can be preferable, taking into account different data trace management and transmission bandwidth requirements.

Section 9.3 reviews some computing platform alternatives that have already been used to support audio processing tasks under a resource-constrained framework.

9.3   AUDIO-ENABLED RESOURCE-CONSTRAINED PLATFORMS

Initially, the first attempts in the design of audio platforms were based on the adaptation of general-purpose hardware devices, for example, those used in the implementation of wireless sensor networks (WSNs). One of the pioneering platforms in this group of devices was the *TelosB*, which was commercially launched in the middle of the last decade and became very popular, owing to its low-power operation and its adaptability to different types of monitoring applications. It has been widely used in a vast number of experimental deployments, and it is still used in applications that do not require a great software complexity, being an example of a low-power platform with limited computational capabilities and transmission bandwidth. The main components of the *TelosB* platform are the *Texas Instruments* MSP430F1611 microcontroller and the *Texas Instruments* CC2420 wireless transceiver. The MSP430F1611 is an ultra-low-power microcontroller with an internal clock of 8 MHz that features 10 KB of *random access memory* (RAM) and 48 KB of flash memory. It is a 16-bit processor with several power-down modes and extremely low sleep current. The MSP430 has eight 12-bit *analog-to-digital converter* (ADC) channels, of which six are accessible on the expansion connector. The ADC input ranges from 0 to 3.0 V, and the maximum sampling frequency is 200 kHz. On the other hand, the CC2420 radio transceiver implements the IEEE802.15.4 standard. It offers reliable wireless communications, with a maximum bandwidth of 250 kbit/s, and a wide range of power management capabilities, with a very low power consumption. Other popular WSN platforms with 8- or 16-bit microprocessors, very similar to the TelosB in terms of power consumption and computational performance are *Mica2*, *MicaZ*, *Iris*, and *Waspmote*. Table 9.1 summarizes the main features of these platforms.

TABLE 9.1
Summary of the Main Features of Popular WSN Motes

Platform	Microcontroller	Transceiver	Memory
Iris Mote	Atmel ATmega 1281	Atmel AT86RF230 802.15.4/ZigBee radio	8 KB RAM + 128 KB flash
Mica2	Atmel ATmega 128L	Chipcon 868/916 MHz	4 KB RAM + 128 KB flash
MicaZ	Atmel ATmega 128	TI CC2420 802.15.4/ZigBee compliant radio	4 KB RAM + 128 KB flash
TelosB	Texas Instruments MSP430	250 kbit/s 2.4 GHz IEEE 802.15.4 Chipcon	10 KB RAM + 48 KB flash ROM
Waspmote	Atmel ATmega 1281	ZigBee/802.15.4/DigiMesh/ RF, 2.4 GHz/868/900 MHz	8 KB SRAM + 128 KB flash ROM

Because these platforms do not contain specific hardware components for audio data acquisition, some companies have released multimedia-sensing boards that can be easily connected to expand the range of available sensors. Some examples are as follows:

- *Tmote Invent*. The hardware components in this platform combine a TelosB sensor module for communication and computation with an integrated suite of sensors (light, accelerometer, temperature, etc.), a microphone, and a speaker. The Tmote Invent microphone circuit allows for omnidirectional acquisition of sounds over a relatively narrow frequency range.
- Easysen WiEye is a low-power flexible sensor board that includes acoustic and low-pass-filtered acoustic signal envelope sensors, specifically designed to be plugged into TelosB devices.
- MTS310 is a flexible sensor board with a variety of sensors, such as dual-axis accelerometer, dual-axis magnetometer, light, temperature, acoustic, and sounder. It can be used with Iris, MicaZ, and Mica2 motes.

Examples of motes and constrained sensor boards can be found in Figure 9.4.

Although it is possible to find hardware elements to build audio acoustic sensor devices with the aforementioned platforms, very few real deployments and applications can be found in the literature. The reduced performance of these microcontrollers and the low memory capacity prevent the execution of a demanding signal-processing application or a high-rate data acquisition. Some exceptions are the works presented in Refs [15,16]. In Ref. [15], it is shown that it is possible to implement a low-complexity method for event detection and localization, based on the efficient cumulative sum algorithm, using the Tmote Invent platform. The authors in Ref. [16] present the results obtained within the *EAR-IT* European project for acoustic monitoring over large-scale smart environments. This work provides a detailed description of the implementation of acoustic monitoring platforms based on the TelosB and the Waspmote nodes, considering their use for real-time acoustic

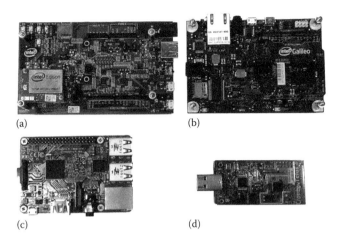

(a)   (b)

(c)   (d)

FIGURE 9.4   (a) Intel Edison. (b) Intel Galileo. (c) Raspberry Pi 2. (d) TelosB mote.

data streaming. The experiments in *EAR-IT* demonstrated the feasibility of performing audio streaming in smart city infrastructures with a relative high audio quality
and a sensor node lifetime of more than 15 h (when the node is constantly capturing
and transmitting the audio signal).

A medium range of audio platforms comprises the proposals presented in Refs
[17,18]. The aim of these works is to provide a tool to visualize audio events in
real time, perform online audio analysis, and save data traces for further offline
processing. To achieve these objectives, the proposed hardware platform, named
Acoustic ENSbox, features a 400 MHz *Intel* PXA 255 microprocessor with 64 MB
of RAM and 32 MB of flash memory. Connected to the processing module, there
are two PCMCIA boards: a sound card with 4 input channels and a Wi-Fi card that
establishes a wireless ad hoc mesh network with the neighboring nodes. The main
drawback of this platform is the high power consumption of its components, which
restricts the node lifetime to only 8 h when it is supplied by batteries.

Recently, a new group of high-range platforms featuring much powerful 32-bit
microprocessors has emerged as an alternative to the developments described above.
This new family of devices is mainly focused on multimedia applications, and these
devices provide an extraordinary capacity of data acquisition, basically because they
are equipped with larger RAMs and *Secure Digital* (SD) flash storage and allow
high-performance data processing. Inside this group of platforms, we can highlight *Intel's Galileo* and *Edison* boards, *Raspberry Pi B+*, *Raspberry Pi 2*, and the
BeagleBoard, as shown in Figure 9.4.

In the cases of the Raspberry and the BeagleBoard platforms, the included hardware enables their use as a complete desktop computer with video output. Another
important issue that affects the management and applicability of these devices is that
all of them allow the use of common general-purpose *operating systems* (OSs). So
far, the most widely OS applied in this field has been *Linux*. In fact, there are specific

Linux distributions focused on the development of embedded systems, such as the *Yocto* distribution used in the Intel platforms, *Noobs* and *Raspbian* for Raspberry, and *Open Embedded* for BeagleBoard. The benefit of using Linux to manage the platform resources is that it enables a fast prototyping of audio applications reusing existing components, drivers, and libraries, such as the *Universal Serial Bus* (USB) audio drivers, the *PortAudio* library, and the modules integrated within the *ALSA* project. In addition, the great effect and popularity that these systems are gaining have attracted the interest of important companies such as *Microsoft*, which has released a *Windows Developer Program* for *Internet of Things*, including Windows versions for Intel and Raspberry platforms. Thus, the OS support enables new possibilities in the development of advanced audio monitoring systems. As in the case of the previous 32-bit architectures, the node autonomy and network lifetime represent the more important limitation when batteries are used to supply the node. Table 9.2 presents the main features of these platforms.

Attention must also be paid to applications exploiting the information from acoustic sensors found in mobile devices. The prevalence of mobile devices, coupled with the proliferation of wireless networks, creates new opportunities for acoustic technology. Mobile devices are small and are used while on the move, not only offering additional possibilities for innovation but also introducing challenges [19].

9.4 APPLICATIONS IN AMBIENT ASSISTED LIVING

Acoustic sensor networks in smart homes can follow different processing strategies. The capabilities of the sensors may range from a simple microphone and transmitter to powerful computing nodes, and special care must be taken in making the trade-off between distributed processing and centralized processing in ad hoc acoustic networks.

Considering simple nodes with severe computational and power limitations (e.g., TelosB), the nodes sample the acoustic signal, perform some simple computation, and send the information to a collector node (database or web server). This collector gathers the information with some web service and processes it with a high-power computational core.

On the other hand, in the case of complex sensing nodes with powerful processors and enough memory to perform complex computations locally, the collecting service can be based on a cloud architecture. For example, an alternative to streaming high-rate data to a collecting node is to extract highly relevant features from the data locally and then upload only the computed features. In this case, it may be desirable to let the nodes send requests for specific processing tasks performed in the cloud, such as training a classifier, performing some analysis on the data, and receiving a suitable energy-aware processing plan [20].

Finally, in cases where the level of processing available may vary from node to node, asymmetric schemes need to be considered. As a general case, a smart home application making use of an acoustic processing network can be assumed to follow the structure shown in Figure 9.5.

Audio-enabled nodes are deployed at different rooms of the house, and they may have the ability to communicate only to a central node or to the rest of the nodes in the

TABLE 9.2

Summary of the Main Features of the Hardware Platforms

	Raspberry B+	Raspberry 2	BeagleBoard X15	Edison	Galileo
Video	HDMI,	HDMI	HDMI	None	None
Audio	Stereo output	Stereo output	Stereo output	I2S Out	I2S Out
Flash	microSD socket	microSD socket	microSD socket; 4 GB onboard flash	SDIO interface; 4 GB onboard flash	microSD socket
RAM	LPDDR 512 MB	LPDDR2 1 GB	DDR3L 2 GB	LPDDR3 1 GB	512 KB on-chip SRAM, and 256 MB DRAM
Processor	Single-core ARM1176JZFS 700 MHz	Quad-core ARM Cortex-A7 900 MHz	Dual-core ARM Cortex-A15 1.5 GHz	Dual-core Atom 500 MHz	Single-core Intel® Quark 400 MHz
Graphics	Broadcom Dual-core VideoCore IV® Multimedia co-processor	Broadcom Dual-core VideoCore IV® Multimedia co-processor	PowerVR Dual-Core SGX544 3D, 532 Mhz	None	None
GPIO	40 pins	40 pins	157 pins	70 pins	26 pin
Ethernet	10/100 onboard	10/100 onboard	10/100/1000 onboard	None	10/100 onboard
WiFi	None	None	None	Dual-band 802.11 (a/b/g/n)	None
Bluetooth	None	None	None	Bluetooth 2.1/4.0	None
USB	4 ports	4 ports	3 USB 3.0, 4 USB 2.0, 1 USB Client	1 USB-OTG	2 ports (AB and B)
Other interfaces	SPI, UART, I2C	SPI, UART, I2C	SPI, UART, I2C, CAN	SPI, UART, I2C, PWM	SPI, UART, I2C, PWM
Power	5 V * 300 mA (~1.5 W)	5 V * 400 mA (~2 W) One-core (+ 150 mA per core)	2 W	3.3–4.5 V at <1 W	5 V * 500 mA (~2.5 W)

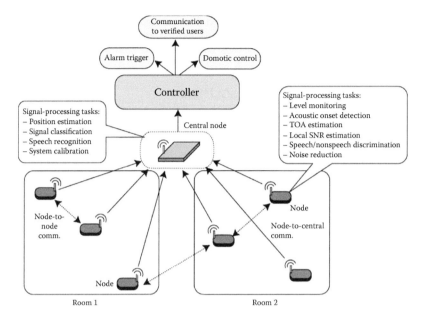

FIGURE 9.5 General structure of acoustic-based AAL system.

network. Depending on the resources of the nodes, they might perform several signal-processing tasks, as described in Figure 9.5. Most of these tasks depend only on the locally acquired signal, although distributed approaches may interchange information from the different nodes to compute other parameter values. Usually, the central node is assumed to be a more powerful node that does not have the restrictions of the rest of nodes in terms of communication bandwidth, power, or computing capabilities. As a result, the most demanding signal-processing tasks are performed at this central node, which usually requires collecting information from the rest of the nodes in the network. Moreover, the central node is usually the one that is directly connected to an intelligent controller, which collects and interprets the data to execute a required action. The controller can send alerts or information messages to emergency services, visual or tactile alerts to a group of verified users, or even perform some domotic action in response to the message received from the central node.

Note that many of the actions required from the system may depend on several aspects. First, it is very important to know when a given sound event has happened with high detection rate (and low false alarm rate). This allows the system to perform a given action only when it is necessary, without interfering or annoying the user significantly. Second, knowing the type of event (baby crying, fall, broken glass, etc.) is also necessary to let the system react accordingly to the detected event. AAD methods allow to determine that some interesting acoustic event might have happened, while AED algorithms are aimed at identifying correctly the type of event. Finally, it may be also very important to have information about the location of the sound source that produced the event. With this information, the system is able to collect more evidence about the type of event or complete the information included in a triggered alert. Section 9.5 discusses these three main aspects of acoustic-based AAL systems.

9.5   SOUND SOURCE LOCALIZATION

To best aid the occupant of a smart home, AAL systems must collect information about him or her. One of the valuable pieces of information is the location of the user. Knowing the position of the user enables the implementation of services and applications that may make the living environment easier, safer, or more comfortable. As an example, the system can make automatic adjustments in electric devices, such as switching off the kitchen oven when the user has been out for some time. In addition, applications can use the positioning data to adjust the lighting conditions of the house or playing music in the rooms where the user is located. Furthermore, real-time location data can be used to detect abnormal activity, providing information to relatives of an elder person who lives alone [21]. The problem of timely and valid notifications of emergency medical services in the context of location-aware systems may be only addressed, provided that an accurate and reliable localization system exists. In fact, first responders' reaction quality can improve considerably in those cases where location's information is available. As a result, AAL platforms may significantly benefit from accurate indoor localization systems.

The *Global Positioning System* (GPS) constitutes a reliable and easily deployable technology in outdoor scenarios. However, GPS is largely unavailable in indoor scenarios because of the large attenuation introduced by building walls and ceilings. For this reason, several systems have already been proposed for indoor localization. Furthermore, indoor environments are strongly affected by multipath propagation as well as non-line-of-sight conditions. A typical smart home environment consists of various rooms, with different kind of furniture and equipment. These conditions strongly affect propagation conditions and can be a real issue for indoor localization systems. Although several systems have been proposed for indoor localization, the available solutions always show a trade-off between localization accuracy and installation complexity (and thus costs).

To perform indoor source localization with *radio frequency* (RF) signals, most systems exploit measurements of physical quantities related to beacon packets exchanged between the nodes and some anchors (devices deployed in the environment whose position is a priori known). Some typical quantities are the *received signal strength* (RSS), the *angle of arrival* (AOA), the *time of arrival* (TOA), and the *time difference of arrival* (TDOA). Since the ranging accuracy depends on both the signal speed and the precision of the TOA measurement, acoustic signals are usually preferred because of their relative low speed. Moreover, the RSS approach using RF signals, while being considerably inexpensive, incurs significant errors due to channel fading, multipath, and in-band signal interferences.

The first approach to TOA estimation requires a time stamp acquisition at the emitter and the receiver to measure the duration of the acoustic signal path. As time capture is performed at the speaker and the microphone during the acoustic signal transmission, time synchronization between the local clocks of these two components must be guaranteed to obtain an accurate TOA measurement. However, the establishment of a common reference clock throughout the whole wireless network is a complicated task. Some indoor location systems that solve synchronization issues by means of RF support are *Bat* [22], *Cricket* [23], *Dolphin* [24], *Cusum-WSN*. [15], and *Teliamade* [25]. They use a combination of RF and ultrasound signals to avoid the problem of requiring a common reference clock. Since RF signals can reach receivers almost immediately in small-range scenarios, its reception is used as a trigger for the acquisition of the initial time stamp at the receiver. Afterward, when the arrival of the acoustic signal is detected, the receiver captures a second time stamp and estimates the TOA subtracting the two marks. Nonetheless, there are still two important uncertainties that affect this estimate: (1) the difficulty in establishing precisely the time of the acoustic signal arrival at the receiver and (2) the extent of the RF transmission delay, which was initially considered negligible. The former can be mitigated by using robust AAD methods. As an example, the cumulative sum approach followed in Ref. [15] has been shown to be useful in detecting speech and impulsive sounds.

In other systems, time synchronization is achieved intrinsically, owing to the system architecture. An example is the sort of systems where all the acoustic nodes are cable-connected to some central equipment that commands and aligns the acoustic signal acquisition. Some instances of systems included in this set are 3D-Locus [26] and the one described in Sertatıl et al. [27].

In some applications, the location of the acoustic nodes is not known in the initial deployment, so that methods able to perform some kind of self-localization of nodes are also of interest. Some of these methods integrate several speakers and microphones that transmit overlapping acoustic signals modulated with different *pseudo-noise* (PN) codes of the same family and present low cross-correlation when different codes are sent at the same time. This feature allows the detection of different PN codes at the receiver, simply correlating the input signal with the sought code, allowing to obtain an estimate of the TOAs for getting the node locations. There is also a proposal to build a ranging system based on TOA of acoustic signals that makes time synchronization needless. This ranging mechanism is called *BeepBeep* [28] and relies on a sender-receiver signal transmission. The scheme was conceived to run on mobile devices; it transmits two audio signals, each sequentially in a different direction. This pair of transmissions allows the determination of the TOA without requiring time synchronization between the two devices.

Typically, a sound source location setup assumes that there is a mobile sound source and a collection of fixed anchors placed at known positions. Once the system anchors estimate the TOA of the acoustic signal emitted by the source, the distances between the source and each anchor can be calculated using the propagation speed of sound. Depending on the localization precision requirements, this step might require an initial calibration process to determine factors that have an strong influence on the speed of sound, such as temperature, relative humidity, pressure, and air turbulence. The computation of the source location can be carried out in a central node by using the estimated distances and applying trilateration. The method is based on the formulation of one equation for each anchor that represents the surface of the sphere centered at its position, with a radius equal to the estimated distance to the sound

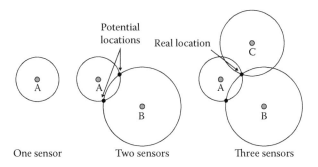

FIGURE 9.6   Localization using trilateration.

source. The solution to this series of equations finds the point where all the spheres intersect. Figure 9.6 shows schematically how at least three sensors are needed to locate a source on a plane. To obtain a three-dimensional localization, one more sensor would be needed. The problem is solved using the geometric trilateration method, but it can be applied only when the distance estimates are free of noise. In practice, when the measurements are affected by noise, there is an uncertainty about the position of the real solution, and the problem can be solved only by employing statistics methods (i.e., least squares and weighted least squares), which find the optimal solution in the resulting areas of uncertainty.

An alternative approach to TOA-based localization is to use TDOAs. Some approaches use several synchronized sensors to estimate the TDOA of acoustic signals by computing a *generalized cross-correlation* (GCC) [29]. The localization principle with TDOAs is similar to trilateration, except that instead of working with spheres, each TDOA measurement defines a half hyperboloid in the three-dimensional space. Figure 9.7 shows how a hyperboloid is defined for each pair of sensors in a localization setup. The source location is the one where all half hyperboloids intersect. Again, TDOAs have usually measurement errors, and most

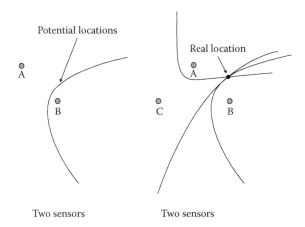

FIGURE 9.7   Source localization using multilateration.

closed-form localization algorithms try to approximate the least-squares criterion. The least-squares problem can be linearized by introducing the source range as an additional variable, leading to the so-called spherical equations. Several approaches have been suggested to solve the linearized problem, such as the unconstrained and constrained least squares [30] and the spherical intersection method [31].

Finally, other types of well-known methods are those based on *steered-response power* (SRP), which have been shown to perform very robustly in adverse acoustic environments. These methods compute the GCC between each pair of sensors, forming a spatial likelihood map by accumulating the GCC values that correspond to the TDOA expected at each spatial position [32]. Despite the significant performance of SRP approaches, they are marginally used in wireless acoustic sensors, owing to their high computational cost and need for synchronization.

9.6 ACOUSTIC ACTIVITY DETECTION

Acoustic activity detection systems are aimed at identifying the temporal boundaries of relevant acoustic events (alarms, shouting, impulsive noises, etc.). Acoustic events appear spontaneously and unpredictably over some background noise, making it difficult to delimit automatically the temporal instants at which these events occur. Knowing with high accuracy when a given sound event has emerged from a background noise is very important for other processing systems that require the event as an input, such as the audio classifiers used in the AED systems, discussed in Section 9.7.

Note that the motivation behind the design of AAD algorithms is that a simple energy threshold on the signal does not work sufficiently well because of several reasons. On the one hand, if the threshold is set too low, the background noise will trigger the detector too often. On the other hand, a high threshold might only be surpassed by high-energy acoustic events, missing many others with moderate amplitude. As an example, consider the sound signal shown in Figure 9.8, where several

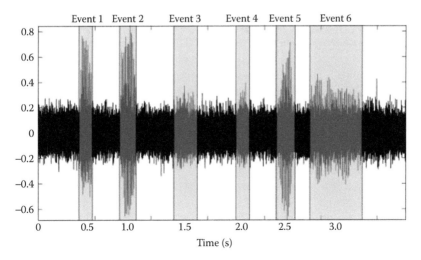

FIGURE 9.8   Acoustic events in the presence of background noise.

events occur in the presence of background noise. While some of the events may be easily detected by an energy threshold, others could not be detected so easily, thus making it necessary to use additional parameters besides the energy. Moreover, the background noise level may change depending on the hour of the day or between day and night conditions. Therefore, adaptive methods are usually preferred.

A set of techniques related to AAD are those aimed at *voice activity detection* (VAD). Voice activity detectors are found in many speech-processing systems, such as telephony systems, speech recognition, and speech transmission over the Internet. The aim of a VAD algorithm is to discriminate between speech and non-speech signal segments. It turns out that detecting speech in background noise is not as trivial as it appears. However, VADs can make use of special features of speech to assist the detection. In this context, the different VAD methods include those based on energy thresholds, pitch detection, spectrum analysis, zero-crossing rate, periodicity measures, higher-order statistics in the linear prediction residual domain, or other combination of features [33,34]. Note that, while having a similar objective as VAD, AAD tries to discriminate general acoustic events from the background and not only the speech, which would result in algorithms making use of different set of features or decision rules than those considered by VAD methods.

Most techniques for AAD are based on two types of algorithms [35]:

- *Metric-based algorithms*: They analyze sound in short-duration temporal frames, calculating a similarity measure (metric) between adjacent regions and comparing this metric with a given threshold. If this measure is above the threshold, it is decided that an event has been produced. Some metrics proposed in the literature are the Euclidean distance [36], the *Mahalanobis* distance [37], the *Kullback-Leibler* distance, and the *Bhattacharyya* distance [38]. In some cases, it is the change in the computed metric that suggests that the onset time of the event, for example, the time derivative of the Mahalanobis distance, has been used for this purpose [37]. In other methods, an adaptive threshold as a function of the average distance over a wider observation window is used [39]. There are other approaches that do not use distance metrics but other parameters such as cross-correlation and energy interpolation [40]. Although their computational cost is not excessively high, they usually need relative long-duration windows (around 4 s) to provide acceptable results.
- *Model-based algorithms*: These methods use model selection techniques such as *Bayesian information criterion* to decide whether the observations correspond to a given model from a set of candidate models [41]. These methods need a long observation window, and therefore, impulsive events are more difficult to detect. Another disadvantage is their high computational cost.

Finally, an alternative approach for AAD would be to consider background noise as a type of event within an AED framework [42]. Section 9.7 describes such kind of systems from a general point of view.

9.7   AUDIO EVENT DETECTION

The possibility of detecting emergency cases in AAL environments is of great importance. Detecting activity or the onset of a given sound event only is not enough for a fully intelligent system. The inability to distinguish between various acoustic events by looking only at the acquired sound levels leads to a significant false alarm rate. It is more reasonable to have embedded acoustic monitoring systems aimed at detecting potentially dangerous situations and to initiate emergency calls when necessary. To tackle this application, AED algorithms are designed to find occurrences of specified target sounds from an input acoustic signal.

Typically, AED algorithms analyze sounds by processing the input signal throughout two different blocks. The first block is usually a feature extractor that extracts the information that is considered to be relevant from an audio signal and that will help the following block in deciding the occurrence of a given type of event. The second block is actually the one that decides whether a target event is present at a given time instant. Figure 9.9 shows these blocks schematically. The input signal is first analyzed by the feature extractor, providing an input feature vector to the classifier that gives the label of the most likely type of event.

Feature extraction in an important stage in audio analysis. In general, it is considered an essential processing step in pattern recognition and machine learning tasks. The extracted features must be informative with respect to the desired properties of the data. Another view of feature extraction is to consider it a data rate reduction procedure, where the original voluminous audio signal is transformed to a more suitable data representation. To achieve this goal, a good knowledge of the application domain is needed, so that only the best features are considered. Acoustic target events can be characterized both in the time domain and in the frequency domain. For example, the discrimination can rely on the harmonic/nonharmonic spectral characteristics of the event by using a proper periodicity feature. On the other hand, temporal characteristics may be useful to discriminate between impulsive and stationary noises. Some very well-known spectral features are the *mel-frequency cepstral coefficients*, which have been widely used in automatic speech recognition for providing a compact representation of the spectrum with auditory considerations. Temporal considerations are also added by using delta features that characterize temporal changes from one analysis frame to another. *Non-negative matrix factorization* (NMF) has also been applied in AED to obtain spectro-temporal features, characterizing both the spectral and temporal structures of the target events [43].

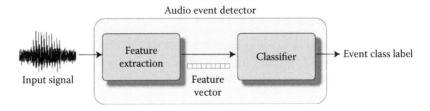

FIGURE 9.9   Main blocks of an audio event detector.

In addition, there are a number of features that have been traditionally used for audio classification tasks [44]:

- *Time-domain audio features*: Short-time energy, zero-crossing rate, entropy of energy, temporal centroid, and auto-correlation.
- *Frequency-domain audio features*: Spectral centroid, spectral spread, spectral entropy, spectral flux, and spectral roll-off.

The second important block in AED is the audio classifier. The classifier is typically based on pattern recognition techniques, which allow to learn the characteristics of an event automatically from a set of training data consisting of examples of the event. Most systems employ established machine learning and pattern recognition techniques such as *Gaussian mixture models* [45], *hidden Markov models* [46], *support vector machines* [43], and *artificial neural networks* [47]. As with the case of the feature selection block, the choice of pattern recognition also depends on the application, since different techniques have different properties in terms of accuracy, noise robustness, and computational complexity. Currently, a lot of research effort is directed at deep-learning techniques [48], which have been shown to provide very good results when a large amount of training data is available.

According to the aforementioned considerations, it becomes quite evident that the target types of events to be detected must be decided at the development stage of an AED system. In this context, the AED system can be designed to detect one or several types of events. For example, in the AAL context, fall detection systems can be of special interest. Smart homes for AAL may also benefit from multiclass systems. An example of an AAL multiclass system is presented in [49], which is aimed at detecting events in elderly healthcare applications. Another multiclass system is the one presented in [50], which implements a detector for 21 different kitchen sounds.

The *IEEE AASP Challenge* addressed the detection of nonspeech isolated events, live events, and synthetic events from an office environment. Several algorithms were proposed and evaluated using common training and test data sets, providing a general view of the state-of-the-art results in AED tasks [51].

9.8 CONCLUSION

Acoustic signal processing realized via WSN can provide interesting solutions when applied in the context of AAL smart homes. Consider, for example, systems where information messages can automatically be triggered and sent to emergency services or relatives in case a fall or similar emergency is detected. This chapter reviewed architectures, hardware platforms, as well as signal-processing algorithms when deploying such systems, also known as WASNs. While especially battery-power and processing constraints are still putting certain limitations on WASNs, signal-processing techniques have made significant progress to allow very accurate detection, classification, and localization of sound events. Hence, we are certain that several of these systems and techniques will form an integral part of tomorrow's AAL smart homes.

REFERENCES

1. United Nations, Department of Economic and Social Affairs, Population Division. *World Population Ageing 2013.* ST/ESA/SER.A/348, 2013.
2. M. Eichelberg, L. Rölker-Denker, A. Helmer and A. Doma, Eds. AAL Joint Programme, Action Aimed at Promoting Standards and Interoperability in the Field of AAL, Deliverable D7: Use Cases in the Ambient Assisted Living Domain: A selected collection from AAL JP, FP6 and FP7 projects. Ambient Assisted Living Association, Brussels, Belgium, 2014.
3. M. Meis, T. Fleuren, E. M. Meyer and W. Heuten. User centered design process of the personal activity and household assistant: Methodology and first results. In *Proceedings of 3rd German AAL Congress.* VDE-Verlag, Berlin, Germany, January 2009.
4. S. Goetze, N. Moritz, J. E. Apell, M. Meis, C. Bartsch and J. Bitzer. Acoustic user interfaces for ambient assisted living technologies. *Informatics for Health and Social Care*, 34(4):161–179, 2010 (SI Ageing & Technology).
5. J. Schroeder, S. Wabnik, P. W. J. van Hengel and S. Goetze. Detection and Classification of Acoustic Events for In-Home Care. In *Ambient Assisted Living*, Springer, Berlin, Germany, pp. 181–195, 2011.
6. M. Crocco, M. Cristani, A. Trucco and V. Murino. Audio surveillance: A systematic review. *ACM Computing Surveys* (CSUR), 48(4):52, May 2016.
7. A. Härmä, M. McKinney and J. Skowronek. Automatic surveillance of the acoustic activity in our living environment. In *Proceedings of the 2005 IEEE International Conference on Multimedia and Expo*, pp. 1–8, 2005. IEEE, Amsterdam, the Netherlands, July 2005.
8. E. Costanza, J. Panchard, G. Zufferey, J. Nembrini, J. Freudiger, J. Huang and J. Hubaux. SensorTune: A mobile auditory interface for DIY wireless sensor networks. In *Proceedings of the SIGCHI Conference on Human Factors in Computing Systems*, pp. 2317–2326, ACM, Atlanta, GA, April 2010.
9. G. Simon and L. Sujbert. Acoustic source localization in sensor networks with low communication bandwith. In *Proceedings of the 2006 International Workshop on Intelligent Solutions in Embedded Systems*, pp. 1–9. IEEE, Vienna, Austria, June 2006.
10. H. Wang, C. Chen, A. Ali, S. Asgari, R. Hudson, K. Yao, D. Estrin and C. Taylor. Acoustic sensor networks for woodpecker localization. In *Proceedings of SPIE 5910, Advanced Signal Processing Algorithms, Architectures, and Implementations XV*, pp. 591009–591009. SPIE, San Diego, CA, July 2005.
11. M. Vacher, P. Chahuara, B. Lecouteux, D. Istrate, F. Portet, T. Joubert, M. Sehili, B. Meillon, N. Bonnefond, S. Fabre, C. Roux and S. Caffiau. The sweet-home project: Audio processing and decision making in smart home to improve well-being and reliance. In *Proceedings of the 35th Annual International Conference of the IEEE Engineering in Medicine and Biology Society (EMBC)*, pp. 7298–7301. IEEE, Osaka, Japan, July 2013.
12. A. Bertrand. Applications and trends in wireless acoustic sensor networks: A signal processing perspective. In *Proceedings of the 18th IEEE Symposium on Communications and Vehicular Technology in the Benelux (SCVT)*, pp. 1–6. IEEE, Ghent, Belgium, November 2011.
13. W. Chen, J. C. Hou and L. Sha. Dynamic Clustering for Acoustic Target Tracking in Wireless Sensor Networks. *IEEE Transactions on Mobile Computing*, 3(3):258–271, 2004.
14. T. Dang, N. Bulusu and W. Hu. Lightweight acoustic classification for cane-toad monitoring. In *Proceeding of the 42nd Asilomar Conference on Signals, Systems and Computers*, pp. 1601–1605. IEEE, Pacific Grove, CA, October 2008.
15. M. Cobos, J. J. Perez-Solano, S. Felici-Castell, J. Segura and J. M. Navarro. Cumulative-sum-based localization of sound events in low-cost wireless acoustic sensor networks. *IEEE/ACM Transactions on Audio, Speech, and Language Processing*, 22(12):1792–1802, December 2014.

16. C. Pham, P. Cousin and A. Carer. Real-Time on-demand multi-hop audio streaming with low-resource sensor motes. In *Proceedings of the IEEE 39th Conference on Local Computer Networks Workshops (LCN Workshops)*, pp. 539–543, IEEE, Edmonton, Canada, September 2014.

17. M. Allen, L. Girod, R. Newton, S. Madden, D. T. Blumstein and D. Estrin. VoxNet: An interactive, rapidly-deployable acoustic monitoring platform. In *Proceedings of the ACM/IEEE International Conference on Information Processing in Sensor Networks. (IPSN '08)*, pp. 371–382, IEEE, St. Louis, MO, April 2008.

18. L. Girod, M. Lukac, V. Trifa and D. Estrin. The design and implementation of a self-calibrating distributed acoustic sensing platform. In *Proceedings of the 4th International conference on Embedded networked sensor systems (SenSys '06)*, pp. 71–84, ACM, Boulder, CO, November 2006.

19. Z. Tan and B. Lindberg, Eds. *Automatic Speech Recognition on Mobile Devices and Over Communication Networks*. Springer, Berlin, Germany, 2008.

20. S. Nirjon, R. F. Duickerson, P. Asare, Q. Li, D. Hong, J. A. Stankovic, P. Hu, G. Shen and X. Jiang. Auditeur: A mobile-cloud service platform for acoustic event detection on smartphones. In *Proceedings of the 11th Annual International Conference on Mobile Systems, Applications, and Services, (MobiSys'13)*, pp. 403–416. ACM, Taipei, Taiwan, June 2013.

21. T. Kivimäki. Technologies for ambient assisted living: Ambient communication and indoor positioning. PhD Thesis. Tampere University of Technology Tampere, Finland, 2015.

22. A. Harter, A. Hopper, P. Steggles, A. Ward and P. Webster. The anatomy of a context-aware application. *Wireless Networks*, 8(2–3):187–197, 2002.

23. N. B. Priyantha, A. Chakraborty, H. Balakrishnan. The cricket location support system. In *Proceedings of the 6th Annual International Conference on Mobile Computing and Networking (MobiCom '00)*, pp. 32–43, ACM, Boston, MA, August 2000.

24. M. Hazas, A. Ward. A high performance privacy-oriented location system. In *Proceedings of the IEEE International Conference on Pervasive Computing and Communications, (PerCom 2003)*, pp. 216–223, IEEE, Fort Worth, TX, March 2003.

25. C. Medina, I. Alvarez, J. C. Segura, A. de la Torre, C. Benitez. TELIAMADE: Ultrasonic indoor location system: Application as a teaching tool. In *Proceedings of the IEEE International Conference on Acoustics, Speech and Signal Processing (ICASSP 2012)*, pp. 2777–2780, IEEE, Kyoto, Japan, March 2012.

26. J. C. Prieto, A. R. Jiménez, J. I. Guevara, J. L. Ealo, F. A. Seco, J. O. Roa, F. Ramos. Performance evaluation of 3D-LOCUS advanced acoustic LPS. *Instrumentation and Measurement, IEEE Transactions on*, 58:2385–2395, 2009.

27. C. Sertatıl, A. Altınkaya, K. Raoof. A novel acoustic indoor localization system employing CDMA. *Digital Signal Processing*, 22(3):506–517, 2012.

28. C. Peng, G. Shen, Y. Zhang, Y. Li, K. Tan. BeepBeep: A high accuracy acoustic ranging system using COTS mobile devices. In *Proceedings of the ACM International Conference on Embedded Networked Sensor Systems (SenSys 07)*, pp. 1–14, ACM, Sydney, Australia, November 2007.

29. C. Knapp and G. C. Carter. The generalized correlation method for estimation of time delay. *Acoustics, Speech and Signal Processing, IEEE Transactions on*, 24(4):320–327, August 1976.

30. P. Stoica and J. Li, Source localization from range-difference measurements. *IEEE Signal Processing Magazine*, 23(6):63–66, IEEE Signal Processing Society, Piscataway, NJ, USA. November 2006.

31. H. C. Schau and A. Z. Robinson. Passive source localization employing intersecting spherical surfaces from time-of-arrival differences. *Acoustics, Speech and Signal Processing, IEEE Transactions on*, 35(8):1223–1225, August 1987.

32. M. Cobos, A. Marti and J. J. Lopez. A modified SRP-PHAT functional for Robust Real-time sound source localization with scalable spatial sampling. *Signal Processing Letters, IEEE*, 18(1):71–74, January 2011.

33. J. Ramírez, J. M. Górriz and J. C. Segura. Voice activity detection: fundamentals and speech recognition system robustness. In *Robust Speech Recognition and Understanding*, Ed., M. Grimm and K. Kroschel, pp. 1–22, I-TECH Education and Publishing, 2007.

34. K. Woo, T. Yang, K. Park and C. Lee. Robust voice activity detection algorithm for estimating noise spectrum. *Electronic Letters*, 36(2):180–181, 2000.

35. A. Temko. Acoustic event detection and classification. PhD Thesis. Universitat Politècnica de Catalunya, Barcelona, Spain, 2007.

36. E. Wold, T. Blum, D. Keislar and J. Wheaton. Content-Based classification, search, and retrieval of audio. *Proceedings of the IEEE Multimedia*, 3(3):27–36, 1996.

37. G. Tzanetakis, P. Cook. Multi-Feature audio segmentation for browsing and annotation. In *Proceedings of the IEEE Workshop on Applications of Signal Processing to Audio and Acoustics*, pp. 103–106. IEEE, New Paltz, NY, October 1999.

38. O. Pietquin, L. Couvreur and P. Couvreur. Applied clustering for automatic speaker based segmentation of audio material. *Belgian Journal of Operations Research, Statistics and Computer Science*, (JORBEL) - Special Issue on OR and Statistics, vol. 41, no. 1–2, pp. 69–81. Bruxelles-SOGESCI, Bruxelles, Belgium, 2001.

39. J. Pinquier and R. André-Obrecht. Jingle detection and identification in audio documents. In *Proceedings of the IEEE International Conference on Acoustics, Speech, and Signal Processing (ICASSP '04)*, pp. iv-329–iv-332, vol 4. IEEE, Montréal, Québec, Canada, May 2004.

40. M. Vacher, D. Istrate, L. Besacier, J. Serignat and E. Castelli. Life sounds extraction and classification in noisy environment. In *Proceedings of the IASTED International Conference on Signal & Image Processing*, IASTED, Hawaï, USA, July 2003.

41. S. Cheng and H. Wang. A sequential metric-based audio segmentation method via the Bayesian information criterion. In *Proceedings of the European Speech Processing Conference (INTERSPEECH)*. International Speech Communication Association (ISCA), Geneva, Switzerland, September 2003.

42. L. Kennedy and D. Ellis. Laughter detection in meetings, NIST meeting recognition workshop. In *Proceedings of the IEEE International Conference on Acoustics, Speech, and Signal Processing (ICASSP '04)*, pp. 118–121. IEEE Montréal, Québec, Canada, May 2004.

43. C. V. Cotton and D. P. W. Ellis. Spectral vs. spectro-temporal features for acoustic event detection. In *Proceedings of the IEEE Workshop on Applications of Signal Processing to Audio and Acoustics (WASPAA)*, pp. 69–72, IEEE, New Paltz, NY, October 2011.

44. T. Giannakopoulos and A. Pikrakis. *Introduction to Audio Analysis*. Elsevier Ltd, Oxford, UK, 2014.

45. L. Besacier, E. Castelli, D. Istrate, J. Serignat and M. Vacher. Smart audio sensor for telemedicine. Smart Object Conference ('2003), pp. 222–225, Grenoble, France, May 2003. https://hal.archives-ouvertes.fr/hal-01085196.

46. T. H. Divakaran, R. Radhakrishnan and Z. Xiong. audio Events Detection Based Highlights Extraction from Baseball, Golf and Soccer Games in a Unified Framework. In *Proceedings of the 2003 IEEE International Conference on Acoustics, Speech, and Signal Processing (ICASSP '03)*, Vol.5, V-632-5. IEEE, Hong Kong, China, April 2003.

47. D. Ellis. Detecting alarm sounds. In *Proceedings of the Workshop on Consistent and Reliable Acoustic Cues (CRAC-2001)*. CRAC, Aalborg, Denmark, September 2001.

48. E. Cakir, T. Heittola, H. Huttunen and T. Virtanen. Multi-label vs. combined single-label sound event detection with deep neural networks. In *Proceedings of the 23rd European Signal Processing Conference (EUSIPCO 2015)*, pp. 2551–2555. IEEE, Nice, France, August 2015.

49. M. Popescu and A. Mahnot. Acoustic fall detection using one-class classifiers. In *Proceedings of the 2009 Annual International Conference of the IEEE Engineering in Medicine and Biology Society*, pp. 3505–3508. IEEE, Minneapolis, MN, September 2009.

50. F. Kraft, R. Malkin, T. Schaaf and A. Waibel. Temporal ICA for classification of acoustic events in a kitchen environment. In *proceedings of European Conference on Speech Communication and Technology (INTERSPEECH)*, pp. 2689–2692. International Speech Communication Association (ISCA), Lisbon, Portugal, September, 2005.

51. D. Giannoulis, E. Benetos, D. Stowell, M. Rossignol, M. Lagrange and M. D. Plumbley. Detection and classification of acoustic scenes and events: An IEEE AASP challenge. In *Proceedings of the 2013 IEEE Workshop on Applications of Signal Processing to Audio and Acoustics*, New Paltz, NY, October 20–23, 2013.

Section III

Case Studies and Field Trials

10 Information and Communication Technology Support for Stroke Care
The Relevance Dilemma

Margreet B. Michel-Verkerke

CONTENTS

10.1 INTRODUCTION

Each year, about 45,000 people in the Netherlands experience a stroke, of which approximately 9,000 people do not survive. The incidence of transient ischemic attacks is estimated at 29,000 people a year. This resulted in approximately 226,000 people suffering from stroke in 2007 in the Netherlands (Nationaal Kompas Volksgezondheid).

For over a decade, Dutch hospitals started having stroke units and started to organize their collaboration with rehabilitation centers, nursing homes, and home care organizations in stroke services. In addition, for other chronic diseases, such as diabetes mellitus and chronic obstructive pulmonary disease (COPD), disease management projects and integrated care initiatives arose. The cross-functional collaboration of care providers is based on shared ideas and knowledge of how to treat the chronic disease. A large body of knowledge is documented in medical guidelines, clinical pathways, and care standards. In the same span of time, electronic health records developed, and at this moment, several vendors in the Netherlands offer information systems supporting integrated care for diabetes, COPD, and cardiovascular risk management. Although 65 stroke services are members of the Knowledge Network for Stroke in the Netherlands (Kennisnetwerk cerebrovasculair accident), no such integrated care information system exists for stroke patients. This research is focused on the reason for this unavailability. What about stroke care is so different from care for other chronic diseases that a stroke care information system is not yet realized?

In Section 10.1.1, the developments in stroke care and the use of information and communication technology (ICT) support of stroke care will be described, followed by a theoretical overview on the success and failure of ICT in healthcare, based on the USE IT-model. This will lead to the research objectives and research question in Section 10.2. Section 10.3 presents the research design and methods. The results can be read in Section 10.4 and will be discussed in Section 10.5. In Section 10.6, the conclusions are drawn.

10.1.1 BACKGROUND

Van der Linden et al. (2001) published an overview of transmural care initiatives in the Netherlands. Transmural care is defined by them, as

> care, attuned to the needs of the patient, provided on the basis of cooperation, and coordination between primary and specialized caregivers, with shared overall responsibility and the specification of delegated responsibilities (Van der Linden et al. 2001).

The word transmural refers to the cross-organizational aspect of the collaboration; literally, transmural means "over the wall," that is, the walls of the hospital. The initiatives for transmural stroke care made up to 4% of the 271 known transmural care facilities at that time. The median number of patients treated in a transmural care facility was 170 (Van der Linden et al. 2001). For stroke care, a national Breakthrough Collaborative Improvement project started in 2002 in the Netherlands. The project included 23 stroke services, which realized significant improvements in stroke care on at least one topic in 83% and on all aims in 34% within 18 months (Minkman et al. 2005). At the beginning of the project, the most frequently mentioned bottleneck in stroke care was length of stay and the second bottleneck was inadequate transfer of information. Other topics were after-care (i.e., care after the medical treatment has ended), thrombolysis treatment, protocols and cooperation, monitoring and management, patient education, and professional expertise (Minkman et al. 2005). After the Breakthrough Collaborative Improvement project ended, stroke services continued to develop and to try to overcome the bottlenecks. Under the slogan "time is brain," especially the acute phase of stroke was improved; more patients arrived in time in a hospital, and thrombolysis was applied more often (Saposnik et al. 2009). Rosendal measured the effect of stroke services on patient outcomes and found no effect on health status. However, the number of patients that needed professional and institutional care after 6 months was reduced. Stroke services seemed to make stroke care more efficient (Rosendal et al. 2002). This seems also the case in the regional stroke service, which is described in this article. Nijmeijer et al. (2005) report that the stroke service reduced the length of stay in the hospital. The stroke rehabilitation unit in the nursing home contributed to the improvement of patient flow.

Transmural care soon developed into so-called "healthcare chain," in which a healthcare chain is defined as

> coherent set of purposeful and planned activities and/or measures, aimed at a specific patient category, phased in time (RVZ 1998, translation by the author).

The word chain refers to the sequential and linear character of the care process, in which each care provider or care organization makes up a link of the chain. Healthcare chain defined in this way was focused on continuity of care. In recent years, healthcare chains became a network character for several chronic diseases, such as diabetes and COPD. Nowadays, healthcare chains resemble the concept of integrated care. Integrated care is defined by Kodner and Spreeuwenberg as

> a coherent set of methods and models of the funding, administrative, organizational, service delivery and clinical levels designed to create connectivity, alignment and collaboration within and between the cure and the care sectors (Kodner and Spreeuwenberg 2002).

For the Dutch situation, the rule for "integral funding" and the presence of a nationally accepted care standard, which steers the collaboration and content of the provided care, enable healthcare chains for diabetes care, COPD care, and cardiovascular risk management to meet the criteria mentioned in Kodner and Spreeuwenberg's definition of integrated care. However, this does not apply to stroke care. Despite the presence of many stroke services in the Netherlands, the national stroke care standard is still a concept and the rule for integral funding does not apply (NZA 2010). Joubert et al. (2009) demonstrated that follow-up care focused on the prevention of a secondary stroke resulted in improvement of risk factors. It is not known how many stroke services include the chronic phase, and if so, whether this brings benefits for the patients.

10.1.2 INFORMATION AND COMMUNICATION TECHNOLOGY IN INTEGRATED CARE

As the implementation of electronic patient records in healthcare progresses, the need and desire for collaborative information systems in integrated care arise. In diabetes care, dedicated integrated care information systems have been successfully introduced and used for several years now and are reported to benefit the patient and the care provider in supporting integrated care (Goshy and Simmons 2006, Featherstone and Keen 2012, Cleveringa 2010). Although differences exist between the various systems, in general, in the Netherlands, these systems are accessible through the Internet and consist of a diabetes patient record, a workflow management system, and a decision support system. The diabetes patient record is filled with information retrieved from other electronic patient records; for example, relevant items from the medical history are retrieved from the electronic medical record of the general practitioner, but care providers also enter data such as actual medication. The workflow management system and decision support system are based on the diabetes care protocol. All care providers involved in the integrated care of the diabetes patient use the diabetes integrated care information system. In the Netherlands, integrated care information systems are also available for COPD, cardiovascular risk management, and thrombosis prevention but not for stroke care.

In stroke care, telemedicine was developed in order to make the expertise of neurologists experienced in performing thrombolysis available to less experienced doctors in rural or remote clinical settings (Hess et al. 2005, Demaerschalk 2011). The research of Mitchell et al. (2011) suggests that it is possible to interpret noncontrast computed tomography brain scans and computed tomography angiogram by reliably using an iPod or iPhone and, by that, facilitate telemedicine for stroke patients in the acute phase.

Despite intense end-user involvement, a custom-made electronic medical record for stroke care in a university hospital did not last after the project phase. The end users saw little benefits, except for more concise reporting. They were also not interested in an isolated system (Van der Meijden 2002, Van der Meijden et al. 2000a, 2000b, 2001, M.J. van der Meijden, 2012, pers. comm.). Hertzum and Simonson's (2008) trial of an electronic patient record at a stroke unit demonstrated a reduction of mental workload and less missing information.

10.1.3 ADOPTION AND ACCEPTANCE OF INFORMATION SYSTEMS IN HEALTHCARE

The research is based on the USE IT model, which theorizes that user characteristics determine adoption. Adoption is defined as

> … adoption, a decision to make full use of an innovation as the best course of action available (Rogers 1995, 21).

USE IT		Domain	
		User	*Information Technology*
Innovation	**Product**	**Relevance** *Macro-relevance* *Definition*: The degree to which the user expects that the IT system will solve his problems or help realize his actually relevant goals. *Micro-relevance* *Definition*: The degree to which IT use helps solve the here-and-now problem of the user in his working process.	**Requirements** *Definition*: The degree to which the user needs are satisfied with the product quality of the innovation. *Macro-requirements* Strategic general requirements and tactical approach is the degree to which the users agree with the objectives and methods used. *Micro-requirements* Functional and performance requirements specify what the content of the innovation should be.
	Process	**Resistance** *Macro-resistance* *Definition*: The degree to which the surroundings and locality negatively influence the users of IT. *Micro-resistance* *Definition*: The degree to which IT users themselves are opposing or postponing the IT change.	**Resources** *Material resources* *Definition*: The degree to which material goods are available to design, operate, and maintain the system. *Immaterial resources* *Definition*: The degree to which immaterial goods are available to design, operate, and maintain the system.

FIGURE 10.1   The USE IT model.

User characteristics are described by four determinants: relevance, requirements, resources, and resistance (Spil et al. 2006). The USE IT model (Figure 10.1) is based on the adoption theory of Rogers (1995), the Technology Acceptance Model (Davis 1989, Venkatesh and Davis 2000), and the Information Systems Success Model (DeLone and McLean 2002).

The four determinants are located on two axes: (1) the innovation axis and (2) the domain axis, to demonstrate that adoption is always affected by the innovation process and the innovation product and, at the same time, always affects both the user domain and the information technology domain. The determinants help identify which characteristics of the user and aspects of the innovation are dominant in a specific case. The USE IT model is suitable for baseline measurement and evaluation purposes (Michel-Verkerke et al. 2006).

Research learned that relevance is the dominant determinant, followed by the requirements determinants (Spil et al. 2006). Adequate resources are a prerequisite for adoption, and resistance is the result of lacking relevance, requirements not met, and lacking resources (Michel-Verkerke et al. 2006). Therefore, the focus in this research will be on the relevance and the requirements determinants of an integrated care information system for stroke care.

10.1.4 Objectives of the Study

The development of integrated care for stroke patients is progressing, but the development of an information system or electronic patient record is lagging behind. In this research, we use the USE IT model to explain why a stroke integrated care information system is not yet adopted, while an integrated diabetes care information system is adopted. This will lead to a more general conclusion about which characteristics of integrated care determine whether a disease integrated care information system will be adopted or not. The research question is: Which characteristics of stroke care determine the adoption of an integrated care information system for stroke services?

10.2 THEORY AND METHODS

10.2.1 Study Context

The regional stroke service had started in 1997 and consisted of the stroke unit of the hospital, the rehabilitation center, a reactivation (i.e., rehabilitation) unit in a nursing home, and a home care organization, all situated in the eastern part of the Netherlands. The geographical regions that are served by each of the participants overlap but differ in size and boundaries. Today, the regional stroke service has expanded and includes another nursing home and organizations for social work, informal care, and an organization to support disabled people. The regional stroke service is a member of the Dutch National Knowledge Network Cerebrovascular Accident.

10.2.2 Study Design

The research has a longitudinal qualitative case study design. A multi-method sociotechnical approach was used, containing both qualitative methods and desk research. The approach was based on Babbie (1995), Cooper and Schindler (1995), and Yin (2009). The semi-structured interviews were based on the USE IT model, which comprises five sections: (1) care process, (2) relevance, (3) requirements (information quality), (4) resources, and (5) resistance (Spil and Schuring 2006). The inclusion criteria for interviewees were: Being involved in stroke care, availability, and being able and willing to express one's opinion and thoughts on the topics.

The research was conducted from 2003 to 2012. The first phase was part of the Freeband Telecare project, which was targeted to improve and optimize the care for people with a stroke, by means of communication and information exchange between the various care providers and institutions by using integrated fixed and mobile ICT applications (Freeband). In this phase (T1), 14 care providers and 2 nurses with management tasks of the regional stroke service were interviewed, using the USE IT model. Between T1 and T2, the developments regarding ICT in stroke care were monitored. At the end of the research (T2), again semi-structured interviews with

TABLE 10.1

Number of Interviewees Per Function of Each Organization

Organization/ Function	Hospital		Rehabilitation Center		Nursing Home		Home/ Primary Care		Vendor Software		External Program Management		N
Phase	T1	T2	T1	T2	T1	T2	T1	T2	T1	T2	T1	T2	
Medical staff	1		1	1	1		1						5
Nurse	1	1	1	1	1		1	1					7
Paramedical staff	1		4	1	1								7
Management/ staff							2	1		1		2	6
n	3	1	6	3	3		4	2		1		2	25

five care providers and four project managers and experts were conducted. Table 10.1 shows the professions and types of care organization of the interviewees.

10.2.3 METHODS FOR DATA ACQUISITION AND MEASUREMENT

The selection of interview candidates was based on the case study's selection strategy of Yin (2009). In the first phase of the research, the stroke service was analyzed and central care provider functions were identified, and of each role, at least one representative was asked to be interviewed. To monitor the developments regarding ICT in stroke care, documents and reports about stroke care in the Netherlands and specifically about the regional stroke service were retrieved from the Internet and from participants in the project; personal communication was also used. The researcher participated in this project from 2008 to 2010.

At T2, key figures in stroke care and related ICT projects were identified and invited to be interviewed. The purpose and procedure of the research was explained by the researcher to each interview candidate, and full anonymity was guaranteed. Interview candidates were free to decide whether they would want to participate or not, without consequences for their position. During the interviews, either notes were taken or the interview was recorded.

10.2.4 METHODS FOR DATA ANALYSIS

To interview care providers, the USE IT model for semi-structured interviews was used (Spil and Schuring 2006). The interviews were analyzed per topic; the answers were broken down in elementary statements and in several iterations aggregated to summarizing statements. The interviews with project managers and experts were summarized to statements and quotes relevant to the research topic. The transcriptions and summaries were sent to the interviewees for comments and approval. During the project, participants were invited to project meetings to hear and discuss research results.

10.3 CASE STUDY RESULTS

10.3.1 STROKE CARE PROCESS

The regional stroke service was initiated in 1997 by the neurologist specialized in stroke care in order to improve stroke care by raising the number of thrombolysis and to create continuity of stroke care. The leading adage was "time = brain." In 2003, stroke care was well organized, as is depicted in Figure 10.2. The regional stroke service protocol served as a guideline for the patient flow and collection of patient data.

As can be seen in Figure 10.2, the stroke service was a linear process. The interviewed care providers stated that cross-organizational cooperation is merely a matter of information transfer than real collaboration. Only the stroke unit organized cross-functional and cross-organizational client meetings, in which the medical specialist of the rehabilitation center participated to decide about the next step in stroke care. After the acute phase, the rehabilitation phase followed, after which the patient went home or was admitted in a nursing home. When asked who was most steering in stroke care, interviewees named the medical specialist in their own organization most frequently, followed by the management and the insurance company. After discharge from the rehabilitation center, the patient was offered four consultations with a nurse during the following 18 months. When a patient came home from the hospital, the home care nurse invited the patient for three consultations during the first year. After this so-called "after-care," the patient left the stroke service, as can be seen in Figure 10.2. The chronic phase was not included in the stroke service. Prevention of a new stroke or other vascular disease was the task of the general practitioner, but this preventive care was not different from prevention of other vascular diseases.

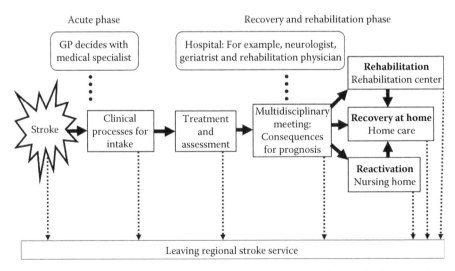

FIGURE 10.2   Patient flow in the regional stroke service, based on freeband-telecare project. Reactivation is rehabilitation in a nursing home.

From 2008 until 2011, care consortia could acquire extra finances for starting a collaborative stroke care program. Four consortia started, and together, they covered the entire region of Twente. A new element in these consortia was the participation of patient organizations, organizations to support informal care, and social work organizations (Menzis 2010). After the ending of the extra finance, the consortia merged to two stroke services—one related to each of the two hospital organizations in the region. The regional stroke service is one of the remaining stroke services and now consists of a hospital, the rehabilitation center, two nursing homes, a home care organization, an organization for informal care, an organization for social work, and an organization for ambulant care of people with disabilities. Not only the stroke service has expanded, but the organization has also changed. The neurologist is still head of the stroke service, but the medical managing director of the rehabilitation center is the president of the steering committee. The two remaining stroke services intend to cooperate, especially concerning the use of ICT. They are developing a shared data set for the exchange of patient information.

From the interviews, it can be learned that, in 2012, the main goal of the stroke service—to reduce the patient's delay and the doctor's delay—has been accomplished. The neurologist is still focused on reducing door to needle time and the length of stay. Financial incentives stimulate this focus. The linear structure of stroke care has not changed. All interviewed care providers state that they exchange information at admission and discharge of a patient, but they do not collaborate across organizations while they provide care. As the vendor states:

> Stroke is not a real chronic disease like diabetes: it starts acute and then you have an end situation. It does not have a chronic discourse with yearly checks and the like, which influence the discourse.

The nurses of the hospital and home care collaborate when organizing events for patients. After-care is still offered either by the rehabilitation center or the home care organization. Not all patients discharged from hospital use the after-care offered by the home care organization. Patients also make little use of the support offered by social work or informal care organizations. However, no specific stroke care is offered after 18 months of discharge from hospital or rehabilitation center. According to the care providers, this is a fallacy, because stroke patients are rather passive and take little initiatives to improve their situation. The patient is only treated by the general practitioner in order to prevent a second stroke. The knowledge about stroke in the chronic phase in primary care should be improved. After several years, health problems are not linked with the stroke anymore.

A regional care standard or guideline still does not exist. Each organization has its own way of providing stroke care. A regional care standard is not only missed for medical reasons but also has a function to steer the collaboration, as the medical specialist of the rehabilitation center states:

> "we should be so robust […] I send a patient to you, but according to a standard, according to a guideline, he should receive this, this and that. Do you offer that? […] I think that you should question each other about that, and I think that in the end a check like that, is necessary." (translation by the author)

In order to raise the level of knowledge about treating stroke patients, the rehabilitation center continues to organize teaching and training on the topic of stroke.

10.3.2 RELEVANCE OF INFORMATION AND COMMUNICATION TECHNOLOGY IN STROKE CARE

In order to measure the relevance of an information system in stroke care, one needs to know what problems the intended users perceive and what they consider important and relevant in performing their tasks. From Table 10.2, it can be learned that in 2003, all care providers were focused on the patient. Contributing to the well-being of patients motivated them to do their job and they wanted to do their job as good as possible (see Table 10.2).

Like in other stroke services (Minkman et al. 2005), the two bottlenecks of the stroke process were the patient flow and the information exchange (see Table 10.2). Nursing homes had a long waiting list for long-stay admissions, which obstructed the discharge of patients from the hospital. A strong desire for a cross-functional and cross-organizational electronic patient record was expressed by many care

---

TABLE 10.2
Relevance: Only Frequencies of 4 and Above Are Listed

What do you experience as important in stroke care?	$n = 16$
Aspects related to the patient focus	12
Aspects related to functioning of the care provider	6
Aspects related to the coordination and improvement of the care process	5
What exceptions or disturbances make that the stroke care or the coordination of stroke care is obstructed?	
Waiting lists for nursing homes "wrong-bed problem"	7
What aspects in being able to do your job do you experience as a bottleneck or as problematic?	
Absence of cross-functional electronic patient record	7
Shortage of staff, either quantitative or qualitative	5
Insufficient computer facilities	4
For which improvements about the entire stroke care process would you really want to make an effort?	
Implementation of a cross-functional electronic patient record	8
Improving the stroke service, especially the patient flow, for example, by making clear agreements	6
After-care project for stroke patients	5
In what way would the use of information and communication technology be of interest to you?	
Replacing the paper record by an electronic record	11
Improving transmural communication	5
Improving communication within the organization	4

providers (see Table 10.2). The interviewees expected that by sharing an electronic record in the entire stroke service, they would have access to all the patient information that they wanted to have. In Section 10.3.3 (requirements), the information needs and the requirements for an electronic patient record are further explored and discussed.

In 2012, the patient flow has been improved: the length of stay at the hospital was reduced from over 10 days in 2003 to 6 days on average in 2011. The bottleneck of waiting lists for long-stay admission to a nursing home has been solved to a great extent by establishing assessment units in the hospital, run by a nursing home. Despite this improvement, the patient flow is still obstructed sometimes, because the capacity of the rehabilitation center and nursing homes is not always sufficient to admit the patient at the moment the hospital wants to discharge the patient.

Another aspect of relevance is the relevance of stroke care for the individual care providers, which varies largely. For the medical and nursing staff providing stroke care in the hospital and rehabilitation center, stroke patients made up 80%–90% of their patient population. However, general practitioners diagnose a new stroke in their practice only once or twice a year. This explains why the general practitioner is not an active member of the stroke care network, which is regretted by most care providers. Moreover, the home care nurses see few stroke patients. Despite the compassion care providers feel, the relevance of stroke care is very low in home care and primary care. According to the care providers specialized in stroke care, the professional knowledge of stroke care and stroke treatment is comparably low in primary care and nursing homes.

Information and communication technology is expected to solve the remaining problem of lacking information in stroke care; however, stroke care and a dedicated stroke care information system are relevant only for the stroke unit and rehabilitation center. A third problem that was mentioned in 2003 is the shortage of staff, in number as well as in educational level. Although shortage of staff is predicted to be an increasing problem, it is not mentioned in the later interviews. It seems that care providers accepted this problem as "the way it is."

10.3.3 REQUIREMENTS

The requirements determinant is focused on information needs and the information process. In Section 10.3.2, information exchange was mentioned as a bottleneck. In this section, this bottleneck is further explored. Most interviewees expressed a broad need for patient information in 2003. Especially, general practitioners need the medical history and medication information in the acute phase. However, this information is not always available (see Table 10.3), as is the problem with other stroke services (Minkman et al. 2005). The patient and his relatives were the most important sources of information.

The neurologist needed information about the patient in later phases to evaluate the effect of the stroke service. The research demonstrated that telephone and letters were the dominant means of transmural communication. In Section 10.3.2, the absence of a cross-functional and cross-organizational electronic patient record

TABLE 10.3

Information Process: Only Frequencies of 4 and Above Are Listed

What patient information do you need to do your job well?	*n = 14*
Medical history, including other diseases and previous treatments	8
Information of physician at admission (e.g., diagnosis)	6
Information about social situation	4
Patient history	4
Information about the stroke (location, progress, CT scan, etc.)	4
Information from other disciplines, from shared patient record, and so on.	4
Information exchange from own discipline	4
What patient information do you miss when providing stroke care?	*n*
Incomplete information at transfers (admission)	8
Information about medical history and comorbidity lacks	5
From whom do you receive patient information?	
Patient	8
Relatives of patient	7
Physician at admission	4
What information do you produce when providing care to stroke patients?	*n*
Patient's situation and recording of own professional acting in own record during the entire care process	10
What information do you give to whom?	*n*
Education and information about care or treatment to patient and relatives	7
Written transmural transfer information for own discipline	7
Result of tests, treatment, and plan to patient and relatives	6
Oral communication within own discipline	4
Other disciplines in own organization	4

is mentioned as a bottleneck, which corresponds to the high score of missing or incomplete information during patient transfers. Care providers do not know what information the next care provider needs.

This problem still needed to be solved in 2012. Several times, researchers or project members made an inventory of who needs what information and when. However, individual care providers still do not know the information needs of their transmural colleagues and do not realize that each care provider and each discipline need different information. Most care providers still use a mono-disciplinary paper patient record and paper forms to exchange patient information. This information exchange can still be improved: the interviewees continue to consider the patient information they receive incomplete, and they continue to suggest to use an information system to share and exchange patient information. A stroke care information system should support the transfer of patient information with each patient transfer and provide each care provider with complete patient information. The medical specialist of the rehabilitation center suggests to make the use of such a system mandatory:

… such a system should supply me with a trigger, a guideline, a control system, on which I can conclude good care and can transfer it to a colleague, and it just should be normal, that I give the information he is entitled to have, before the patient is on his doorstep.

For physicians, the discharge letter serves as a summary of the record, which covers their own information needs and not necessarily the information needs of the general practitioner, who is the intended receiver of the letter. When a general practitioner works at the central doctor's post during shifts, he can request a "professional summary" of the patient's general practitioner's information system (for all patients and not just for stroke patients). This professional summary provides the general practitioner with actual medication information, actual medical problems, and relevant information from the patient's medical history. By this system, their information needs during shifts are satisfied.

10.3.4 RESOURCES AND RESISTANCE

Almost all respondents expressed a positive attitude toward the use of ICT in healthcare, although they expected some resistance, caused by lack of computer self-efficacy with some of their colleagues. The obstacles the care providers saw in the technical prerequisites seemed to have been resolved in 2012. Now, all care providers are better equipped with computers and internet access.

10.3.5 DEVELOPMENT OF AN INTEGRATED STROKE CARE INFORMATION SYSTEM

In order to start at the beginning of the stroke care process, the Freeband Telecare project resulted in 2004 in the design of a personal digital assistant (PDA) for general practitioners, which they could consult when visiting a possible stroke patient and which would send the general practitioner's diagnosis and a minimal medical data set to the neurologist in the hospital. The use of the PDA was not continued after the end of the project, because general practitioners did not want a solution just for stroke (Michel-Verkerke et al. 2006). As a result of the National Breakthrough Project Stroke, a paper-based transmural patient record was implemented in 2004. This was a patient-held record—apart from the care provider's patient record—that the patient was supposed to present to each care provider in the stroke service. In practice, patients did not always bring the record and care providers did not always write in it. That is why, the transmural stroke record was cancelled after a few years.

New initiatives were taken to develop a stroke integrated care information system. However, a feasibility study in 2006 learned that stroke care was not enough to establish a positive business case. Combination with an information system for mediation in transfers from hospital to follow-up care would create a business case, although only a part of the stroke patients needed this mediation. Nursing homes and home care organizations were interested in this system when the system would also contain patient information needed by nurses. Otherwise, the system would have little relevance for them. This e-transfer information system is still being developed.

Moreover, in other regions of the Netherlands where initiatives were taken to start a stroke integrated care information system, none of them succeeded. One electronic health record for stroke exists, but this is a general e-transfer information system, extended with a stroke form. According to a vendor, a commercial stroke care information system is not realistic as long as there is no national care standard to build it on. Another problem was raised by one of the program managers: none of the care providers feels responsible for the entire stroke care process. A "director" or owner of the entire stroke care process is missing. It is also not clear what the purpose and benefits would be of a stroke care information system.

> those kind of things help, of course, very much in this field. Like: is finance available? Is anyone responsible? Is there anyone you could call a contractor for that care-product. And the market is the stroke patient. All those kind of things are not there. (translation by the author)

In 2006, the project to realize an e-transfer record has started. After several changes, a project organization was established in 2010, comprising a steering committee and workgroups. All care organizations that provide stroke care participate, including organizations for informal care and social work. This project organization covers more organizations than the members of the regional stroke service. A general patient information transfer system is being developed, which can be used by the transfer office of the hospital and all care providers, and not just by the participants in stroke care. The system will have forms per discipline, with additional sections for specific types of care, such as stroke care. Physicians will also be able to share information in this system, but the focus will be on the information needs of the transfer office and nurses, since physicians already have possibilities of electronic exchange of patient information. The transmural information exchange between physicians is already supported by ICT, as far as it concerns formal communication, such as referral and discharge letters.

10.3.6 NATIONAL DEVELOPMENTS IN STROKE CARE

Nationally, a knowledge network for stroke was established in 2005, with the purpose of stimulating the further improvement of care for stroke patients. Stroke services can become a member in order to exchange and raise the knowledge about cerebrovascular accidents. The concept version of the national care standard for stroke was published in November 2011 (Kennisnetwerk CVA 2011). The care standard describes specifically what should be done in the acute phase and gives advices for the chronic phase but does not supply a work flow structure, such as in diabetes care, where periodic checkups are incorporated in the protocol (Cleveringa 2010). The stroke care standard confirms that the stroke care process is linear: from the acute phase to the chronic phase. For the chronic phase, a central care provider is advised. Regarding information processes, the need for quality indicators and the need for information exchange (transfer) are expressed. There seems no need for a stroke integrated care information system (Concept Zorgstandaard CVA/TIA).

10.4   DISCUSSION

10.4.1   WHICH CHARACTERISTICS OF STROKE CARE DETERMINE THE ADOPTION OF AN INTEGRATED STROKE CARE INFORMATION SYSTEM?

From the research, it can be learnt that stroke care does not meet the criteria for integrated care for several reasons:

1. The care process is mainly linear and sequential, with transfers of patient and information, but with little cross-functional and even less cross-organizational collaboration during delivering care.
2. The stroke service covers only the acute phase, the rehabilitation phase, and the first year and a half of the chronic phase and does not cover the entire period for which a patient suffers from stroke.
3. The national care standard is not yet final and does not cover all phases.
4. Because of this, stroke care does not meet the criteria for integral funding in the Netherlands.
5. A great variety in relevance of stroke care exists for care providers; the relevance in primary care is very low.

It is unlikely that stroke care will meet the criteria for integrated care and integral funding in the future. This implies that the chance that a stroke integrated care information system will ever be adopted is very low. The most feasible way to solve the problem of adequate patient information transfer in the stroke service is probably the approach chosen by the stroke services in this case study: to create an e-transfer information system, with additional forms for stroke care.

10.4.2   RESULTS IN RELATION TO OTHER STUDIES

Görlitz and Rashid (2012) designed an architecture for postacute stroke management. A central role in the stroke care that they describe is that of the stroke manager, who organizes the care for the stroke patient after discharge from the clinic in order to guarantee continuity of care. This is comparable with the "transfer office" task, as described in the case study. The stroke management architecture is based on a customer relation management system. Through the architecture, patient information is retrieved from the electronic patient record and distributed to the stroke manager, patient, and next care providers. A stroke health book is added in order to provide the patient with relevant information and to enable the patient to upload measurements. It seems that after discharge of the patient, all care providers will use their own electronic records. It is not clear whether stroke health book still has a function for the patient or care providers in the chronic phase. The design seems to resemble the idea of an e-transfer system (Görlitz and Rashid 2012).

The diabetes decision support system is not only accepted and adopted but also successful in reducing cardiovascular risks in diabetes type-2 patients (Cleveringa 2010). One reason for the success is that the system is based on the national diabetes stroke protocol and supports care providers in following the protocol, since a workflow management system and a decision support system are incorporated. In this way, the main function of the system is to support care providers in following the protocol and to support them in the collaboration needed according to the protocol, as well as to provide the patient with information for this.

Stroke care does not have a national protocol yet. The concept does not cover the entire process and does not steer cross-functional and cross-organizational collaborations. Neither do repeated consultations or checkups take place. Moreover, the incidence and prevalence of stroke (45,000 and 226,000, respectively, in 2007) are much lower than the incidence and prevalence of diabetes (71,000 and 740,000, respectively, in 2007), so the relevance of stroke is much lower for the individual care provider than the relevance of diabetes (Nationaal Kompas Volksgezondheid). In addition, the relevance of a dedicated solution for stroke care is relevant for only a few care providers. The prevention of stroke and other vascular diseases on the other hand is very relevant, which explains the increase of the use of integrated care information systems for cardiovascular risk management (Nationaal Kompas Volksgezondheid).

10.4.3 STRENGTHS AND WEAKNESSES OF THE STUDY

The qualitative research design of interviews bears the risk of subjectivity and the limitation of not being able to justify the outcomes. However, the present research did not intend to justify facts but to explain an existing situation and to distil rules from reality in order to contribute to theory. The longitudinal design is a strong design, since the chance of reacting to incidents or "fashionable" features is low.

10.4.4 MEANING AND GENERALIZATION OF THE STUDY

The burden of the disease of a stroke is second highest after coronary diseases, mainly caused by the long period of time a person lives with the disabling consequences of the stroke (Nationaal Kompas Volksgezondheid). Prevention of disabilities by applying thrombolysis is therefore an important improvement achieved by many stroke services. However, attention to disabling consequences after the patient returns to home is almost as important. The research demonstrates that a stroke service does not cover the entire stroke care process but usually ends about one year after the patient returns home. The patient is lost/out of sight, and later health problems are not related to the previous stroke anymore. The stroke care process can be characterized as a "one-way route," with information exchange, but little communication, and even less collaboration. Transferring information from one care provider to the other should be supported by ICT. Regarding the number of stroke patients, the relevance of stroke care is high for care providers in the stroke unit and the rehabilitation center but rather low for general practitioners and home care. For patients, however, the relevance of adequate and timely started care is very high. In the acute phase, the stroke service seems to function well, but in the chronic phase, it is not clear whether stroke patients receive adequate care. Information and communication technology can support the communication in stroke care, but a stroke integrated care information system is not relevant to all care providers, and this bears the risk of not being adopted. However, a general patient e-transfer information system is highly relevant to most care providers, except to general practitioners, since they already have several electronic services. For them, adoption of connection to their electronic medical records will enhance adoption.

10.4.5   UNANSWERED AND NEW QUESTIONS

This study demonstrates that the chances for a stroke integrated care information system to be adopted are low, because stroke care does not meet the criteria for integrated care and also because the relevance of stroke care for care providers in primary care is low. It would be interesting to investigate whether the development of stroke care in the chronic phase is medically relevant for patients, as is suggested by several interviewees. If so, the question arises: What will be the effect of development of chronic stroke care on the characteristics of stroke care, with respect to the criteria for integrated care and the relevance of stroke care in primary care? Another question that is worth investigating is: Will the suggested solution of an e-transfer-system will solve the information problem in stroke care?

10.5   CONCLUSION

According to most interviewed care providers, the way to solve their information problem is to create a cross-organizational and cross-functional electronic record for stroke patients. However, the general practitioners were very clear: they would not use a solution dedicated to stroke, since they treat very few stroke patients and they do not want a specific solution for each single health problem. This holds for all care providers, except for the stroke unit and the rehabilitation center. That is why, the project in Twente focuses on a general e-transfer information system, with additional sections or forms for stroke care.

When looking at the acute phase, important progress has been made. More stroke patients are treated with thrombolysis, and the patient flow has improved. In addition, the possibilities for stroke care in the chronic phase have increased, but patients do not seem to have discovered them. No regional stroke care protocol or standard has been developed, but a concept of the national care standard has been published recently. It is likely that this national care standard will be implemented by the participating care organizations.

According to the definition of Kodner and Spreeuwenberg, stroke care is not integrated care, since stroke care is a linear process with patient and information transfers, but without real collaboration and no shared funding, and the chronic phase is not covered. In addition, a shared stroke protocol that covers all phases of stroke care and that guides all care providers does not exist.

Care providers feel that there is a strong incentive to improve the quality of care, because they observe that not all stroke patients receive the care they need. To improve the quality of care, cross-functional and cross-organizational collaboration is necessary. Obstacle for this close collaboration is the gap in relevance between care providers and, related to this, the great variation in expertise. A national or regional accepted stroke care standard could help overcome the lack of expertise in primary care, on the condition that the most recent standard is available at the moment and location where it is needed. Another reason why an integrated stroke care information system is not likely to be adopted is the linear and sequential character of the stroke service. The main purpose of the collaboration is to create continuity of care and not integrated care. An e-transfer information system would be able to improve the availability of patient information at the moment and location where it is needed, especially when specific forms for stroke are included.

10.6 COMPETING INTERESTS

From 2008 to 2010, the author was employed at the program office IZIT as a project manager for the e-transfer information system. Information from this period is used in this article. The author does not have any interest in the development of the information system or the stroke service.

ACKNOWLEDGMENTS

The first phase of the research was part of the Freeband Telecare project and was granted with a subsidy of € 175,000. The author would like to thank Ton Spil and Roel Schuring for their contribution to the first phase of the research.

REFERENCES

Babbie, E. 1995. *The Practice of Social Research*. Belmont, CA: Wadsworth Publishing Company.

Cleveringa, F. 2010. *Computerized Decision Support Task Delegation and Feedback on Performance in type 2 Diabetes Care: The Diabetes Care Protocol*. Utrecht, the Netherlands: University of Utrecht.

Cooper, D.R. and P.S. Schindler. 1995. *Business Research Methods*. Singapore: McGraw-Hill.

Davis, F.D. 1989. Perceived Usefulness, Perceived Ease of Use, and User Acceptance of Information Technology. *MIS Quarterly* 13:319–340.

Demaerschalk, B.M. 2011. Seamless Integrated Stroke Telemedicine Systems of Care: A Potential Solution for Acute Stroke Care Delivery Delays and Inefficiencies. *Stroke* 42:1507–1508.

DeLone, W.H. and E.R. McLean. 2002. Information Systems Success Revisited. In *35th Hawaii International Conference on System Sciences*, 7–10 January 2002, Big Island, Hawaii.

Featherstone, I. and J. Keen. 2012. Do Integrated Record Systems Lead to Integrated Services? An Observational Study of a Multi-Professional System in a Diabetes Service. *International Journal of Medical Informatics* 81:45–52.

Freeband-Telecare project. http://www.freeband.nl/kennisimpuls/projecten/telecare/ENindex.html (accessed June 19, 2012).

Görlitz, R. and A. Rashid. 2012. Stroke Management as a Service—A Distributed and Mobile Architecture for Post-Acute Stroke Management. In *European Conference Information Systems,* 10–13 June 2012, Barcelona, Spain.

Goshy, G. and D. Simmons. 2006. Diabetes Information Systems: A Rapidly Emerging Support for Diabetes Surveillance and Care. *Diabetes Technology & Therapeutics* 8:587–597.

Hertzum, M. and J. Simonsen. 2008. Positive Effects of Electronic Patient Records on Three Clinical Activities. *International Journal of Medical Informatics* 77:809–817.

Hess, D.C., S. Wang, W. Hamilton, et al. 2005. REACH, Clinical Feasibility of a Rural Telestroke Network. *Stroke* 36:2018–2020.

Joubert, J., C. Reid, D. Barton, et al. 2009. Integrated Care Improves Risk-Factor Modification after Stroke: Initial Results of the Integrated Care for the Reduction of Secondary Stroke Model. *Journal of Neurology, Neurosurgery & Psychiatry* 80:279–284.

Kennisnetwerk CVA. 2011. *Concept Zorgstandaard CVA/TIA* [concept care standard Stroke – TIA]. http://www.kennisnetwerkcva.nl/concept-zorgstandaard-cva-tia [in Dutch] (accessed June 19, 2012).

Kennisnetwerk cerebrovasculair accident Nederland [Knowledge network cerebrovascular accident Netherlands]. http://www.kennisnetwerkcva.nl [in Dutch] (accessed August 19, 2016).

Kodner, D.L. and C. Spreeuwenberg. 2002. Integrated Care: Meaning, Logic, Applications, and Implications—A Discussion Paper. *International Journal of Integrated Care* 2(November):1–6. http://www.ijic.org/index.php/ijic/article/view/URN%3ANBN%3ANL%3AUI%3A10-1-100309/134 (accessed August 19, 2016).

van der Linden, B.A., C. Spreeuwenberg, and A.J.P. Schrijvers. 2001. Integration of Care in The Netherlands: The Development of Transmural Care Since 1994. *Health Policy* 55:111–120.

van der Meijden, M.J. 2002. *An Electronic Patient Record for Stroke: Development, Implementation and Evaluation in Practice.* Maastricht, the Netherland: Universiteit Maastricht.

van der Meijden, M.J., H. Tange, J. Troost, and A. Hasman. 2001. Development and Implementation of an EPR: How to Encourage the User. *International Journal of Medical Informatics* 64:173–185.

van der Meijden, M.J., H. Tange, J. Boiten, J. Troost, and A. Hasman. 2000a. An Experimental Electronic Patient Record for Stroke Patients. Part 1: System Analysis. *International Journal of Medical Informatics* 58–59:111–125.

van der Meijden, M.J., H. Tange, J. Boiten, J. Troost, and A. Hasman. 2000b. An Experimental Electronic Patient Record for Stroke Patients. Part 2: System Description. *International Journal of Medical Informatics* 58–59:127–140.

Menzis. 2010. *Projecten Ketenzorg VV Regio Twente CVA 28-04-2010* [Projects Integrated care Primary Care Region Twente Stroke]. http://www.menziszorgkantoor.nl/web/Zorgaanbieders/Zorginkoop/KetenzorgProjecten.htm [in Dutch] (accessed June 19, 2012).

Michel-Verkerke, M.B., R.W. Schuring, and T.A.M. Spil. 2006. The USE IT Model Case Studies: IT Perceptions in the Multiple Sclerosis, Rheumatism and Stroke Healthcare Chains. In *E-health Systems Diffusion and Use: The Innovation, the User and the USE IT Model*, ed. T.A.M. Spil and R.W. Schuring, Chapter X. Hershey, PA: Idea Group Publishing.

Minkman, M.M.N., L.M.T. Schouten, R. Huisman, and P.T. van Splunteren. March 23, 2005. Integrated Care for Patients with a Stroke in the Netherlands: Results and Experiences from a National Breakthrough Collaborative Improvement Project. *International Journal of Integrated Care*, 5. http://www.ijic.org (accessed August 19, 2016).

Mitchell, J.R., R. Sharma, J. Modi, M. Simpson, M. Thomas, M.D. Hill, and M. Goyal. April–June 2011. A Smartphone Client-Server Teleradiology System for Primary Diagnosis of Acute Stroke. *Journal of Medical Internet Research* 13(2). http://www.jmir.org/2011/2/e31/ (accessed August 19, 2016).

Nationaal Kompas Volksgezondheid [National Public Health Compass]. http://www.nationaalkompas.nl/ [in Dutch] (accessed August 19, 2016).

Nederlandse Zorg Autoriteit [Dutch Care Authority]. 2010. *Uitvoeringstoets Uitbreiding integrale bekostiging ketenzorg, verruiming grenzen en soorten ketens.* [Performance test expansion integral funding integrated care, expansion limits and types of care]. Nederlandse Zorg Autoriteit. http://www.nza.nl/104107/105832/139003/Uitvoeringstoets-Uitbreiding-integrale-bekostiging-ketenzorg.pdf [in Dutch] (accessed August 19, 2016).

Nijmeijer, N.M., B.M.A.D. Stegge, S.U. Zuidema, H.J.W.A. Sips, and P.J.A.M. Brouwers. 2005. Effectiviteit van afspraken binnen de Regionale stroke-service om patiënten met een beroerte adequaat te verwijzen van de stroke-unit in het ziekenhuis naar een verpleeghuis voor kortdurende reactivering. [Effectiveness of agreements within the Enschedese stroke service in order to adequately refer patients with a stroke from the stroke unit in the hospital to a nursing home for short-term rehabilitation]. *Nederlands Tijdschrift voor Geneeskunde* 149:2344–2349 [in Dutch].

Raad voor de Volksgezonheid en Zorg. 1998. [Council of Public Health and Care]. *Redesign van de eerste lijn in transmuraal perspectief.* [Redesing of primary care in transmural perspective], Raad voor de Volksgezondheid en Zorg. http://rvz.net/uploads/docs/Redesign_van_de_eerste_lijn_in_transmuraal_perspectief.pdf (accessed June 19, 2012 [in Dutch]).

Rogers, E.M. 1995. *Diffusions of Innovations.* New York: The Free Press.

Rosendal, H., C.A.M. Wolters, G.H.M. Beusmans, L. de Witte, J. Boiten, and H.F.J.M. Crebolder. March 2002. Stroke Service in the Netherlands: An Exploratory Study on Effectiveness, Patient Satisfaction and Utilisation of Healthcare. *International Journal of Integrated Care* 2(1):1–9. http://www.ijic.org/index.php/ijic/article/view/50/100 (accessed August 19, 2016).

Saposnik, G., M.K. Kapral, S.B. Coutts, J. Fang, A.M. Demchuck, and M.D. Hill. 2009. Do All Age Groups Benefit From Organized Inpatient Stroke Care? *Stroke* 40:3321–3327.

Spil, T.A.M. and R.W. Schuring. 2006. USE IT Interview Protocol. In *E-health Systems Diffusion and Use: The Innovation, the User and the USE IT Model*, ed. T.A.M. Spil and R.W. Schuring, Chapter XI. Hershey, PA: Idea Group Publishing.

Spil, T.A.M., R.W. Schuring, and M.B. Michel-Verkerke. 2006. USE IT: The Theoretical Framework Tested on an Electronic Prescription System for General Practitioners. In *E-health Systems Diffusion and Use: The Innovation, the User and the USE IT Model*, ed. T.A.M. Spil and R.W. Schuring, Chapter IX. Hershey, PA: Idea Group Publishing.

Venkatesh, V. and F.D. Davis. 2000. A Theoretical Extension of the Technology Acceptance Model: Four Longitudinal Field Studies. *Management Science* 46:186–204.

Yin, R.K. 2009. Applied Social Research Methods Series. *In Case Study Research: Design and Methods*, ed. L. Bickman and D.J. Rog. Vol. 5. Thousand Oaks, CA: Sage Publications.

11 INCA Software as a Service
Sustainably Delivering Cloud-Based Integrated Care

Lars T. Berger, Miguel Alborg Dominguez,
Beatriz Martinez-Lozano Aranaga,
and Vicente Peñalver Camps

CONTENTS

11.1   INTRODUCTION

11.1.1   SOCIO-SANITARY CARE

Healthcare services have been the pride of European democracies, but they have not evolved to respond to the modern environment and are no longer fit for the purpose.[*] While continuing to be based on the common values of universality, access to good quality care, equity, and solidarity, healthcare services must accommodate new realities and acknowledge the need for cost-efficient investments. Social care has traditionally a smaller budget than healthcare,[†] which is why the most urgent need for cost savings is felt in the healthcare system.

There is now an increasing movement of policy makers, medical personal, and social care givers, agreeing that cost cuttings on the basis of resource optimization, prevention (pro-action vs. reaction), and an increased shift to out-patient treatment (care at home) are necessary and can be achieved under the overarching concept of *integrated socio-sanitary care*.[‡] An interesting overview at *European Union* (EU) level is provided, for example, in the "Compilation of Good Practices" by the Action Group B3.[§]

Integrated socio-sanitary care is still in its infancy, and a plurality of definitions exists. To add to the confusion, other terms such as *continuum of care*, *coordination of care*, *discharge planning*, *case management*, and *seamless care* are often used synonymously. It is most frequently equated with *managed care* in the USA, *shared care* in the UK, *trans-mural care* in the Netherlands, and other widely recognized formulations such as *comprehensive care* and *disease management*. Despite, integrated socio-sanitary care is seen as the way forward that will benefit all Europeans (in particular older people), while helping to address resource efficiency

---

[*] This, ultimately, is the fundamental conclusion of the EU Task Force report, *Redesigning health in Europe for 2020*. (European Commission, eHealth Task Force Report—Redesigning health in Europe for 2020, Luxembourg: Publications Office of the European Union, 2012, doi:10.2759/82687, available online http://ec.europa.eu/newsroom/dae/document.cfm?doc_id=2650, last accessed 23/06/2016).

[†] Organisation for Economic Co-operation and Development (OECD), *Health at a Glance 2013: OECD Indicators*, Nov. 2013, OECD Publishing, http://dx.doi.org/10.1787/health_glance-2013-en, available online http://www.oecd.org/els/health-systems/Health-at-a-Glance-2013.pdf, last accessed 23/06/2016.

[‡] North West London, Whole System Integrate Care, web portal http://integration.healthiernorthwestlondon.nhs.uk/, last accessed 23/06/2016 Care.

[§] European Commission—European Innovation Partnership on Active and Healthy Ageing—Action Group B3 on replicating and tutoring integrated care for chronic diseases, A Compilation of Good Practices, 2nd Ed., Nov. 2013, available online, http://ec.europa.eu/research/innovation-union/pdf/active-healthy-ageing/gp_b3.pdf, last accessed 23/06/2016.

and sustainability of care systems. We define it as "collaboration, alignment, training and connectivity among social care, healthcare, and community care providers with the mission to provide better services at reduced cost."

Some concepts also include informal caregivers into these integration tasks. The objective is to achieve integration between hospital, community home, and self-care. Therewith, integrated socio-sanitary care is located at the interception of two traditional markets, the health service market and the social service market, as indicated in Figure 11.1.

Integration frequently leads to multidisciplinary networks consisting of care professionals and informal care givers, which tailor their service provisioning in form of patient-centric models to the patients' care requirements. An example of such a patient-centric multidisciplinary care network is provided in Figure 11.2.

This patient-centric "Circle of Care" is adopted by the EU-funded INCA project (http://www.in3ca.eu), which is successfully running pilots in Spain, Cyprus, Latvia, and Croatia. To enable a practical top-level management, patients are segmented and associated with standardized care plans, also called *care pathways*, which are then adapted by the case manager to customized *integrated care plans* (ICPs) to cater to each patient's individual needs.

11.1.2 PATIENT SEGMENTS AND MARKET DIMENSIONS

Driven by broad shifts in demographics and disease status, long-term conditions absorb by far the largest, and growing, share of healthcare budgets. Over 100 million citizens, or 40% of the population in Europe above the age of 15 years, are reported to have a chronic disease, and two out of three people who have reached retirement age have at least two chronic conditions.[*] Moreover, it is widely acknowledged that

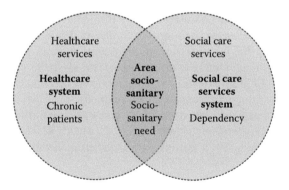

FIGURE 11.1   Integrated socio-sanitary care.

[*] Communication from the Commission to the European Parliament, the Council, the European Economic and Social Committee and the Committee of the Regions, eHealth Action Plan 2012-2020 – Innovative healthcare for the 21st century, Brussels, 6.12.2012 COM(2012) 736 final, available online http://eur-lex.europa.eu/legal-content/EN/TXT/PDF/?uri=CELEX:52012DC0736&from=EN, last accessed 23/06/2016.

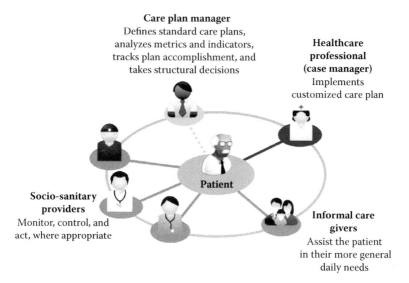

FIGURE 11.2   Patient-centric, multidisciplinary integrated care network.

70% or more of healthcare costs are spent on chronic diseases. This corresponds to more than €700 billion in the EU and is expected to rise in the coming years.[*] As this is putting pressure on the sustainability of health and social care systems, and on the wider economy and society, many integrated-care deployments focus on high-risk chronic patients. The idea is that, by focusing on a relatively small group of only around 30% of the chronic population, which in the current systems are associated with 70%–80% of costs, significant cost savings can be achieved.

The current systems are unsustainable in the medium term owing to demographic and lifestyle changes, the impact of chronic diseases, and budget limitations.[†] Extrapolating the 10% cost-saving prediction made in NWL/NHS2014[‡] to the healthcare spending in the INCA pilot countries leads to the saving potations, as outlined in Table 11.1. It gives a glimpse of the *return on investment (ROI)* that might be obtained by investing into integrated-care solutions.

"The (US) healthcare integration market is expected to grow at a *compound annual growth rate* (CAGR) of 9.6% in the forecast period, to reach $2,745.9 million by 2018 from $1,737.3 million in 2013. Factors such as the rising healthcare costs, presence

---

[*] European Union Health Policy Forum, Answer to DG SANCO consultation on chronic diseases, 13 January 2012, available online http://ec.europa.eu/health/interest_groups/docs/euhpf_answer_consultation_jan2012_en.pdf, last accessed 23/06/2016.

[†] Directorate-General for Internal Policies, Policy Department A—Economic and Scientific Policy, Workshop on "e-Health", Meeting Document, 24 September 2013, available online http://www.europarl.europa.eu/document/activities/cont/201309/20130920ATT71712/20130920ATT71712EN.pdf, last accessed 23/06/2016.

[‡] National Health System (NHS), North West London Integrated Care Pilot: Business Case, Report, London, UK, 2014, available online http://www.cipd.co.uk/pm/b/mainfeed.aspx?UserID=6359, last accessed 23/06/2016.

TABLE 11.1
Savings Potential Assuming a 10% of Healthcare Cost Reduction Due to the Introduction of Integrated Socio-Sanitary Care

Pilot and Reference Countries	Healthcare Spending (2011)[a] (€)	Integrated Care Savings Potential Based on 10% Savings Assumptions (€)
Spain	98,859,560,000	9,885,956,000
Cypros	1,006,060,000	100,606,000
Croacia	3,218,640,000	321,864,000
Latvia	1,266,820,000	126,682,000
EU 27	1,045,171,070,000	104,517,107,000
United States	1,919,270,270,000	191,927,027,000

[a] Eurostat, Expenditure of selected healthcare functions by providers of healthcare, data from 2011 except Cyprus and Latvia which are from 2008 and 2009, available online http://appsso.eurostat. ec.europa.eu/nui/show.do?dataset=hlth_sha1m&lang=en, last accessed 23/06/2016.

of strong government support and initiatives, growing need to integrate healthcare systems, and high returns on investments have increased the demand for healthcare integration. However, various interoperability issues, presence of a fragmented end-users market, and the high cost of implementation of healthcare integration are the factors that are restraining the growth of this market to a certain extent.

North America accounted for the largest share of 65%–70% of the global healthcare integration market, followed by Europe, with a share of nearly 20%. However, the Asian countries represent the fastest-growing markets. The high growth in these countries can be attributed to the increasing awareness regarding healthcare, growth in healthcare spending in emerging countries, and the presence of a large and diverse population in this region."[*]

Apart from reducing financial stress on health and social care systems, it is also expected that with the introduction of socio-sanitary care quality improves due to the following:

- "Stricter adherence by all health professionals to evidence-based care protocols used across multiple organizations
- The provision of high-quality services outside of hospital
- Pro-active care to ensure that long-term conditions do not deteriorate and patients do not need acute care
- Greater support for self-care

[*] Markets and Markets, *Healthcare Integration Market by Product (Interface Engine, Medical Device Integration, Media Integration), by Service (Implementation, Support, Training), by Application (Hospital, Laboratory, Radiology, Clinics) – Global Forecast to 2021*, Market Report, April 2016, available for purchase online http://www.marketsandmarkets.com/Market-Reports/healthcare-it-integration-market-228536178.html, last accessed 23/06/2016

• Increased involvement in their care planning, with multidisciplinary teams drawn from the various constituent organizations of the ICP"*

On the other hand, integrated care is also expected to improve the working conditions and experience of the professional caregivers, owing to the following:

• "Involvement in development of evidence-based care protocols for use across multiple organizations. The integrated care pilot allows all clinicians and care professionals the opportunity to develop protocols to be applied by their colleagues from other settings.
• Greater development opportunities across multiple settings and organizations. All professionals in the integrated care pilot will benefit from direct input through multidisciplinary groups and other opportunities for creating real-time support from their colleagues.
• Access to better (more and improved quality) information about their patients' care by implementing improved information flows between providers, allowing each one to access the most up-to-date records regarding patients in their care."†

Despite all these advantages, integrated care for patients with complex needs and long-term illness is currently not implemented to its full potential, leading citizens with long-term conditions to have a lower quality of life.‡

11.1.3 POLITICAL SUPPORT OF INTEGRATED CARE

The integration of sociosanitary healthcare has been on the European states' agendas for more than 20 years, but despite the introduction of various policies and strategies on national and EU level, and the existence of many initiatives and projects in different European regions, there is still much to be done. The ground is now fertile for significant changes. Although support across Europe is currently greater than ever, and we could say it is gaining "momentum," owing to the difficult integration (many different stakeholders involved), reaching the objectives will take time. The technologies and techniques are not the main barrier but the necessary human, professional, and system changes. The major issue is that the nature of the problem is sociopolitical and not technological, as there are separate legal and funding structures between healthcare and social care. As the number of older citizens rises all over the European countries, at the same time, this translates to higher numbers of patients that face moderate chronic health diseases and lack of accessible family support and thus are at risk of compromised health. These people are suffering from

---

* National Health System (NHS), North West London Integrated Care Pilot: Business Case, Report, London, UK, 2014, available online http://www.cipd.co.uk/pm/b/mainfeed.aspx?UserID=6359, last accessed 23/06/2016.
† Ibid.
‡ N. Goodwin, A. Dixon, G. Anderson, W. Wodchis, Providing integrated care for older people with complex needs—Lessons from seven international case studies, The King's Fund, Jan. 2014, UK, available online http://www.kingsfund.org.uk/sites/files/kf/field/field_publication_file/providing-integrated-care-for-older-people-with-complex-needs-kingsfund-jan14.pdf, last accessed 23/06/2016.

more chronic ailments that require sustained intensive care. There is also a growing need for better, personalized services from their health providers. The European society has made many efforts and created many services to help these citizens, but these services are still segmented into many organizational clusters such as health, social care, housing, and others. There is lack of coordination among these clusters, and their offered services are separately organized and delivered (and the same goes for their respective staff). Healthcare leaders around Europe are seeking to improve the quality of healthcare delivery and broaden access to basic services for the demanding population. At the same time, they are trying to keep costs low. Most European countries and especially the European South (Italy, Spain, Portugal, and Greece) are being hit from the financial crisis, and as a result, their healthcare systems are under similar financial pressures and are facing many new challenges in meeting increased demands for care.

The lack of coordination among the care stakeholders brings with it miscommunications among the various involved parties, contradictory legislations, duplication of procedures, and inefficiencies. As a result, the patients are unhappy with their care, where the unhappiness stems from relatively high costs and accessibility problems. Many countries are trying to take measures to eliminate these inefficiencies and to promote integration through the establishment of strong regulations at local and national levels to promote coordination and collaboration. However, some governments—apparently mainly because of national culture—delay measures for the promotion of integrated care, and if they finally decide to do it, they just take soft and inadequate actions.[*] As a result, policy makers are under less pressure to develop legislative measures and structures of formal integrated care. Thus, the change is easiest to happen in countries with long tradition of integrated care policy such as the Nordic countries. Nonetheless, there is no guarantee for a successful implementation of measures and policies even when governments succeed in creating legislation for their promotion. For instance, decentralization may slow down decision-making processes. In addition, the presence of too many decision-making processes and decision makers can be an obstacle to change.

The experiences in many European countries such as Finland, Sweden, Netherlands, and England demonstrate that clear legislation for integrated care and appropriate financial incentives encourage stakeholders to establish integrated care arrangements.[†] Moreover, the position and the different roles of the several actors within the institutional framework, where the health and social systems operate, define the formation and the implementation of respective policies. Although, in all countries, the central government has the final responsibility for implementing and coordinating health and social care services, the position of its power differs from country to country. In some countries (e.g., UK, Greece, and Cyprus), the central government has strong decisional power, while in others (e.g., in Sweden, Spain, Finland, and the Netherlands), the system is characterized by decentralized powers. In addition, the actors that play a vital role in the decision process differ per case,

---

[*] I. Mur-Veeman, A. van Raak, A. Paulus, Comparing integrated care policy in Europe: Does policy matter?, Elsevier Journal of Health Policy, Volume 85, Issue 2, February 2008, pp. 172–183, doi:10.1016/j.healthpol.2007.07.008, available online, http://www.sciencedirect.com/science/article/pii/S0168851007001728, last accessed 23/06/2016.

[†] Ibid.

and we observe that there are countries (e.g., Finland, Sweden, and Denmark) where municipalities/regions play the main role, while in others, the power lies in the hands of providers—professionals and nongovernment organizations. For example, we could refer to the case of Austria, where the hospitals and medical specialists have a prominent role in the decision-making regarding integrated care services. In the Swedish system, regions along with the municipalities develop a pivotal role in the process of policymaking and care delivery.* In addition, the most essential characteristic of the Dutch healthcare system is that important operative decisions are taken, to a remarkably high degree, jointly by medical professionals and patients.

The way of promoting integrated care is also of importance. In some cases, the central government uses mandatory legislations and obligatory rules with more hierarchical directions, while in other cases (that could be characterized, mostly decentralized), policy makers mostly produce supporting policy notes, recommendations, and guidelines. Local and regional authorities may have more impact on care delivery if they own healthcare centers, primary care services, or nursing homes.

11.1.4 OVERVIEW OF INCA PILOTS

IN3CA (the name coming from "INclusive INtroduction of INtegrated Care," and in short, INCA) has started a pragmatic initial deployment in Europe. For this, five pilot sites are implemented: two in Spain and the others in Cyprus, Latvia, and Croatia. After having completed the implementation tasks, pilots are run for more than a year, followed by an evaluation to validate the implementation of the model and its impact, as well as its market replication potential in other countries.

The various INCA pilots are departing from very different levels of integration, with a plurality of political and organizational obstacles to be overcome. However, INCA does not want to be directly engaged in the titanic task of removing systemic silos, which might take decades to change. Instead, INCA aims to overcome silos, creating a virtual integration in a wise and pragmatic way, in order to leverage the benefits of ICTs, even when other barriers still remain.

INCA with its partnership and holistic approach can help create good and effective communication and coordination channels across the entire care process. Scaling up and generation of critical mass at EU level are a key for successful implementation but require overcoming operational silos, fostering new organizational changes, innovative business models and incentive measures, convergence of technology, and promotion of standards toward interoperable ICT tools.

INCA, therefore, acknowledges that these important *patient-centered holistic inclusive eHealthcare* resources need to be integrated and sustained within regional and local programs and initiatives, beyond the factual silos, "hidden to users" by friendly and smarter INCA interoperability resources, so that better coordination

---

* I. Mur-Veeman, A. van Raak, A. Paulus, Comparing integrated care policy in Europe: Does policy matter?, Elsevier Journal of Health Policy, Volume 85, Issue 2, February 2008, pp. 172–183, doi:10.1016/j.healthpol.2007.07.008, available online, http://www.sciencedirect.com/science/article/pii/S0168851007001728, last accessed 23/06/2016

and integration of service delivery can be smoothly achieved, reusing previous investments in health and social care.

INCA is specifically targeting on the micro level the *service integration* that has a direct impact on the patient's experience and on the standard working practices of care providers and allows INCA to get started right away, with no need to wait for higher-level integration to happen.

11.1.5 Section Summary and Chapter Outline

European demographic trends demand cost cuttings on the basis of resource optimization, prevention (*pro-action vs. reaction*), and an increased shift to out-patient treatment (*care at home*). It is believed that this change can be brought about through the overarching concept of *integrated socio-sanitary care.*

INCA deploys a multichannel, patient-centered, integrated socio-sanitary care platform. Social services, medical organizations, patients, and private care givers are able to interact with each other through any device capable of running an Internet browser. Serving content from the Cloud allows access anywhere at any time.

Apart from the INCA pilots, numerous other integrated care pilots are emerging, such as *epSOS*, *Integrated Home Care*, and *People2People*. An especially visible set of pilots is run with support of the UK National Health Services in the North West London area. INCA is keeping a close eye on these developments, on the one hand to learn and on the other hand to assist and contribute its own experiences.

The various INCA pilots are departing from very different levels of integration, with a plurality of political and organizational obstacles to be overcome. Unfortunately, health and social care systems in Europe are very diverse, making it much harder for service providers to replicate their services. Even within a country, structures and entities are nonhomogeneous, establishing a serious entry barrier for smaller players and SMEs. For our first rollouts, it is of utmost importance to understand existing health and social system care structures, as well as existing market actors in each pilot country.

The following sections present the individual pilots in more detail, putting especial emphasis on the question whether INCA costs are outpaced by INCA cost savings and therewith assuring financial sustainability of the service offering.

11.2 THE SPANISH PILOT QUART/MANISES

Before INCA, Manises hospital—located in the region of Valencia in Spain—already saw the need to change its care delivery model, and when INCA came across with its software platform ADSUM+, it was clear that Manises hospital would join the INCA consortium. Besides, when checking out solutions offered by competing providers, it turns out that INCA/ADSUM+ is very competitive not only technologically but also from a price point of view, which confirms the suitability of the direction taken.

The organizational change that started in 2012 was initially supposed to involve only clinical hospital staff. However, when entering the INCA project, Manises hospital realized that involving primary care providers, who are under management of the hospital in 14 municipalities, including Quart de Poblet (a neighboring city of Manises), would enhance coordination and pilot success.

To go ahead, in order to define new ways of delivering care in an integrated manor, the *Multi-Disciplinary Group* (MDG) was launched. Within this group, different clinical profiles from primary care and acute care levels discuss and agree on how Manises hospital's integrated care pathways should look like. Family doctors, nurses, clinical social workers, specialty doctors, "at home" care unit providers, and medical directors participate in the MDG group. In parallel, a preliminary pilot group to test how to deploy new integrated care pathways was created. Quart de Poblet Primary Care Centre was chosen as the pilot facility to develop Manises hospital's chronic care agenda.

In Spain, social services constitute a protection system to serve the welfare of all people, in particular those who require special attention because of their advanced age, those who have a disability, or because they are at risk of social exclusion. Until now, competences of local councils (in cities with over 20,000 inhabitants) have worked in isolation from other areas affecting citizens directly, as are health services.

Social workers at Quart's Social Services Department evaluate, track, and attend social needs and resources for Quart's population. For doing their work, Quart's Social Services Department has used its own "social pathway," where issues such as caregivers' support and personal autonomy are measured. Before the INCA pilot, "social work" was performed without access to relevant information on chronic patients (such as patients with heart failure [HF]), and there was a high risk to miss candidates eligible for receiving social services and resources. The same happened at Manises hospital, knowing nothing (or near to nothing) about the social situation of the patients they take care of, which could result in an "inefficient usage" of health services (patients may not be ready to stay at home).

Quart's Department of Social Service often offers a last hope for disadvantaged residents. How could Quart better connect citizens to services that deliver the greatest impact and lasting outcomes?

Our data sources are nonclinical and unstructured—such as level of dependency, social status, exercise activity, social interactions, and so on—and can provide a more holistic picture of an individual, potentially making these data better predictors of future health issues.

On both sides (clinical side/social side), we agree that data should no longer be confined to the relatively limited set of data that each side is able to capture. Both parts can benefit if working in a better coordinated way. Overall, the first beneficiary will be the patient.

During the INCA project, Quart is transforming how it serves clients along the social care continuum. Solutions offered in INCA help Quart assess citizens' needs, match citizens to services and benefits, and track progress and outcomes. INCA also helps Quart to integrate with the clinical services provided by Manises hospital.

In the INCA pilot, Quart de Poblet Social Workers are joining Manises hospital's MDG to share a common integrated care pathway when attending the needs of chronic patients with HF from Quart de Poblet. This integration results in a new network of available resources to:

- Effectively exchange nonrestricted clinical and social information for better decisions on how care is to be delivered
- Better reach patients, so that no potential candidate for pro-active care models, delivering clinical and social care, is excluded or forgotten from this new HF care pathway
- Improve chronic patients' perception of quality of services
- Manage more efficiently economic resources at Manises Health Department and also at Quart de Poblet's Local Council

Working with the care center professionals who attend our citizens to develop and progress approaches to integrated care in order to address the coordination and provision of services for patients in full understanding of local factors is proving to be a great experience. Clinical and social information flows among providers, taking into account whatever privacy restrictions and consent forms patients need to sign, allowing socio-sanitary providers to make better care delivery decisions.

When looking at sustainability in financial terms, it is important to address the questions of *where does the money come from*, *where does it go*, and *are there any profits generated through these transactions*. These questions are addressed for the Manises/Quart pilot as follows.

11.2.1 WHERE DOES THE MONEY COME FROM?

Regions healthcare system financing in Spain follows the allocation formula, based on a per capita criterion, weighted by population structure, dispersion, extension, and insularity of the territory. A 2009 revision modified the per capita criterion, shifting to population adjusted by effective health-protected population, population of school age and of those aged 65 years and above, plus the previous geographical factors.

Manises hospital receives a fixed yearly amount of money per capita for all its residents and associated municipalities. The payment is independent of the amount of treatments received. Further, if a nonresident receives treatments, Manises hospital bills the hospital (or private insurance) responsible for this patient. The amount billed is higher than the actual cost produced. If a patient associated with Manises hospital is receiving treatment in another hospital, this other hospital bills Manises (also most likely a price that is higher than the cost that would have been incurred if the patient had been treated in Manises). Hence, it is profitable for Manises to attract patients from other hospitals, and it is desirable that patients associated with Manises receive treatment in Manises and not in other hospitals. It is clear that in such a model, patients are clients, and loyalty of these clients to Manises hospital is a business objective. To achieve this loyalty, Manises hospital strives to offer exceptional services and has some famous doctors/surgeons among its staff that attract clients even from far away. Further, if patients that are associated with Manises hospital are healthy, they make a low use of hospital resources. In this case, there is a profit in the per capita money received for healthy patients. Hence, to maintain the population associated with Manises hospital as healthy as possible is also a business objective.

11.2.2 Savings Prediction Due to the Implementation of HF Pathways

INCA has started to be first applied to the population of patients with HF. The variable costs that patients with these pathologies are producing have been broken down in Table 11.2.

It is expected that INCA can reduce the number of hospitalizations by 20% and the visits to the emergency services by 10%. On the contrary, visits to dependent health centers are increasing by 20% (indicated by the minus sign in Table 11.2). Before starting INCA, these numbers were estimated based on an extensive literature review. Very shortly, we will have evidence for these numbers collected through INCA's internal data gathering, as well as cross-connections to Manises's data analytics system.

Should the predictions be confirmed, then savings due to implementing the HF pathway with the help of INCA will turn out to be € 216,475 annually, as pointed out in Table 11.2.

TABLE 11.2
Cost for Intervention and INCA Relates Savings Considering Heart Failure (HF)

Concept	Cost Per Resource Usage (€/day)	Heart Failure Related Resource Usage (2011)	Heart Failure Related Cost (2011) (€)	Cost Reduction Due to Chronic Care Model Implementation (%)	Cost Reduction Due to Chronic Care Model Implementation (€)
Hospitalization in Manises	122,80	9007	1,106,059,60	20	221,211,92
Hospitalization billed to other hospitals	267,70	188	50,327,60	20	10,065,52
Visit to Manises emergency services	60,70	2090	126,863,00	10	12,686,30
Visit to Manises emergency services billed to other hospitals	131,30	153	20,088,90	10	2,008,89
Visit to a Manises dependent health center	22,70	5658	128,436,60	−20	−25,687,32
Visit to a Manises dependent health center billed to other hospitals	38,80	491	19,050,80	−20	−3,810,16
Total			1,450,826,50		216,475,15

11.2.3   HF Pathway Implementation and Maintenance Costs

Besides the pure software-related costs, consisting of license, adaptation, and maintenance costs, there are significant costs to adapt the internal processes of Manises hospital to implement HF pathways, as detailed in Table 11.3.

11.2.4   Payback of HF Pathway Implementation

Figure 11.3 shows the two different levels of costs, as well as the savings in cumulative values over 24 months.

TABLE 11.3

Costs Prediction in Manises Considering Heart Failure (HF)

Concept 1	Amount (€)	Frequency
	Product Costs	
License	60,000	One off payment
Maintenance fee	12,000	Yearly
Customization	6,000	One off payment
	Adoption Cost	
IT Infrastructure	1,800	One off payment
Organizational changes	56,000	One off payment
Duplicated efforts	20,000	One off payment
Others	4,000	One off payment
Training	3,200	One off payment
Other costs	3,000	One off payment

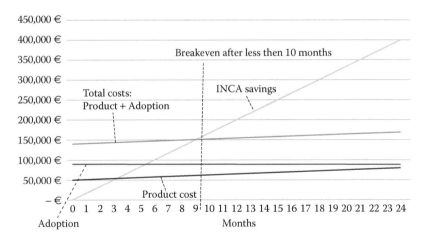

FIGURE 11.3   Costs and savings due to INCA in the case of Heart Failure in the Manises/Quart Pilot.

It can be seen that even with a conservative estimate, the implementation of the HF pathway with the help of the software tool INCA/ADSUM+ is reaching break-even point before 10 months of operation and is producing annual profits afterward.

11.2.5 Concluding Remarks on the Spanish Pilot Quart/Manises

In summary, streamlined processes and integrated information help to speed the delivery of vital benefits. Teams make decisions based on real insight, not instinct, driving better client outcomes. Manises hospital, but also the Social Services Department of Quart Council, has implemented detailed care delivery plans to deliver integrated care over the next years. Ongoing funding is contingent on successful delivery of implementation plans, with clear accountability for performance and expenditure.

Considering that INCA can be rolled out over other pathologies such as chronic obstructive pulmonary disease, stroke, asthma, and diabetes, the profits will scale, bearing in mind that initial kick-off problems have already been overcome through the HF pilot. Profits will multiply by the number of pathways adopted.

It can be concluded that INCA is an economical sustainable solution for Manises hospital, with significant quality-improving side effects for our employees, the patients, and their relatives. This has the potential to further boost Manises hospital's good image in the Valencian Community and beyond, which in consequence can help maintain or even boost patient inflow (patient loyalty), with positive financial effects that have not even been considered in this sustainability analysis.

Also for Quart, the pilot seems to be a very satisfactory approach, and if successful on the long run, it will be scaled up and rolled out across the rest of the health areas of Manises hospital, where the experience of Quart will serve as an example for other city councils to follow.

11.3 THE SPANISH PILOT MURCIA

The Murcia Health Service has been trying to normalize the care of the main health problems through agreed guidelines in different areas, having three objectives: the continuity of care; the reduction in clinical variability, with all their consequences on patient safety; and finally, an increase in efficiency. However, until INCA appeared, the objectives were reached with limited success, partly owing to a lack of a solid organizational software tool.

To demonstrate the impact of INCA, patients with diabetes and HF have been chosen for running the pilot with the objective of improving their care by

- Doing an appropriate stratification, establishing clinical pathways agreed between all stakeholders of the chain of care
- Reconstructing (if needed) the entire history of the patient, backed by the INCA tool that allows visibility (clinical and social) according to the permits granted to each role
- Monitoring performance and tracking patients' evolution with the INCA tool

In summary, Murcia Pilot is trying a better integration of health and social care, with the aim to improve patients' conditions, decrease duplication of tasks, reduce errors, improve prognosis and quality of life, and reduce the frequency of hospitalization.

However, the way to INCA implementation in the Murcia Pilot has not been an easy one. We have had to overcome several barriers as

- Data protection in clinic and social information
- Problems to share the information between health and social departments, even when health and social competencies are owned by the same authority, as in the case of Murcia region
- Split social competencies over different entities and territorial levels
- Short-term view of decision makers responsible for budget allocation
- Social-side low technology knowledge in the elderly and in the social groups that are the objective of INCA

We are glad to say that these barriers have been overcome by now, with the pilot operating successfully now.

11.3.1 MURCIA PILOT SUSTAINABILITY NUMBERS

Data collection of the Murcia INCA pilot is still at a very early stage, which is why the following sustainability analysis is based partly on estimations. Savings and costs of product and adoption have been summarized in Table 11.4.

Plotted over a period of 24 months, it can be seen that savings quickly significantly outpace INCA acquisition and maintenance costs, as well as the costs for INCA adoption (Figure 11.4).

TABLE 11.4
Costs and Savings Prediction in the Murcia Pilot

Concept	Amount (€)	Frequency
Product Costs		
License	200,000	One off payment
Maintenance fee	36,000	Yearly
Customization	30,000	One off payment
Adoption Cost and Management Overhead		
IT Infrastructure	134,800	One off payment
IT Hosting license	1,800	Yearly
Coordination-gp	66,000	Yearly
Case management nurses	588,000	Yearly
Administrative staff	420,000	Yearly
Program dev. and implementation	163,307	One off payment
Savings	1,889,274	Yearly

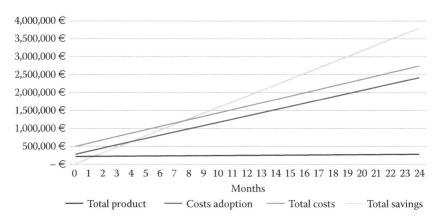

FIGURE 11.4   Costs and savings due to INCA in the Murcia Pilot.

A break-even point is reached after less than 8 months of operation, with significant profits generated from month 9 onward. When comparing with the Quart/Manises pilot, one has to bear in mind that the sample of Quart/Manises is significantly lower (31,066) than the one of Murcia (216,451). This means that Murcia is approximately 7 times bigger, which is reflected by higher costs for change management/adoption, license fees, and maintenance.

11.3.2   Concluding Remarks on the Spanish Pilot Murcia

It can be concluded that INCA can be sustainably run in the region of Murcia, after the end of the pilot phase. We are confident that our data analytics will soon be able to confirm the estimations, and we are confident to have found a solution that fulfils our needs in the case of integrated care, while at the same time being a cost-effective solution.

The new regional government shows great commitment with service continuation and clearly sees INCA as strategic.

11.4   THE CYPRUS PILOT

11.4.1   Introduction of Geroskipou/Cyprus Pilot

Before INCA, there were no pre-existent systems in the Geroskipou Social Care Center. Thus, all the medical data were paper-based. A doctor visited the social care center every week to examine and monitor the health condition of around 60–65 patients per visit (250 patients per month) and advised the social caregivers of what physical exercise, medical treatment, and special diet the patients should follow.

All the patient data were written down and updated through paper. The Social Care Center has about 200 registered members, and on average, it serves 60 persons daily. The doctor and the social caregivers had to face data duplication and loss of data in many cases, and their efficiency was insufficient.

As there were no pre-existent systems in the Social Care Center, the INCA platform came as a standalone solution for our pilot and a total new experience for the people involved, opening new ways of using of IT tools in the socio-sanitary practice, as well as an empowerment of end users. The INCA system has already been translated into Greek, and the training for the pilot members, who are playing the roles of social workers, caregivers, and updaters, has been completed.

Thus, INCA provides the opportunity for Geroskipou Municipality to use a digital tool to monitor the evolution of elder patients suffering from cardio-vascular diseases, while at the same time making the monitoring more efficient and effective.

All the patients have been registered in the platform and assigned to a doctor and a caregiver. Through the platform, they have been also assigned to care plans and care actions. At each visit, the doctor contacts them, monitors their status, and updates their information in the platform. Moreover, some officers of Geroskipou Municipality are collaborating with the doctor toward the healthcare provision and working partly on the program, playing the roles of updater, program manager, social caregiver, and so on. IASIS hospital supports the pilot from the clinical side, and the company INTERFUSION supports the pilot from the technological side.

11.4.2 EXPECTATIONS OF GEROSKIPOU MUNICIPALITY

Geroskipou Municipality expects that:

- INCA will eliminate the information in the form of hard copies, helping to convert all the patient data into electronic form.
- The proposed development can increase the efficiency and the responsiveness of the municipality, resulting in time savings, generation of revenues, and, in the long run (in 5-year period), generation of working positions.
- INCA will help increase digital literacy of all patients of the area, who until now have little or no ICT knowledge.
- Moreover, Geroskipou Municipality plans to use INCA for the new medical center that is going to open in Geroskipou in 2016 (with an overall budget of € 1.5 million) and will support all the territory of not only Geroskipou but also the Paphos District.

11.4.3 OUTSIDE BARRIERS

The main barrier in the Cyprus market is the economic/budgetary issue, as most public administrations and hospitals are experiencing budget cuts and are reluctant to invest on something new. In addition, it is not easy to find sponsoring for such kind of services.

Moreover, there is a technological barrier. The elderly in Cyprus have inadequate technological and IT knowledge, and most of them are not willing to even try to learn or use such kind of systems.

Furthermore, one of INCA's main challenges is organizational. Attitude and change of mind are the main barriers when introducing new care models. Most health professionals and managers continue to defend the service models they have grown up with and consider these models the only way of delivering safe health and social care. Much of the medical staff believes that the parameters of their working model are the only proven and safe ways to provide their service. As a consequence to this reluctance to change, there is no integrated health system and there is very low usage of IT in medical centers, in general.

11.4.4 Analysis of Costs/Benefits

- Reduction of total yearly costs for providing service: manpower and operating costs
- Social care givers would enhance their knowledge of cardio-vascular diseases and strokes, something that could lead to identifying symptoms at an earlier stage and increasing their skills and customer care
- Cost savings due to reduction of:
  - Unnecessary visits of healthcare staff
  - Time saved in internal queries by staff
  - Time saved by the patients
  - Time saved in internal processes
- Significant improvement in quality of life of the target group
- Improving and simplifying the services delivered
- Greater engagement
- New collaborations
- Reduced duplication (many patients have to be registered again and again, as there was no electronic file)
- Upgradation of the already-offered services
- Promotion of e-participation
- Strengthening our bonds with our citizens
- Improvement of the proximity with citizens
- Improvement of the communication between civil servants and citizens
- Deeper impact on the population and especially on the "difficult-to-reach" audience such as the elders and people with impairments
- Improved image of a responsive and efficient administration held by society

11.4.5 Cyprus Pilot Sustainability in Numbers

Savings as well as costs of product and adoption have been summarized in Table 11.5.

Table 11.5 shows the product costs, consisting of license, maintenance fee, and customization. The numbers presented assume 15,000 patients. Further, it shows the adoption costs, consisting of IT Infrastructure, organizational changes, training, and others costs. Finally, it shows the yearly savings, consisting of the sum of indirect revenues, and eliminates costs of the previous care program.

Plotted over the period of 24 months in Figure 11.5, it can be seen that the savings are initially not enough to counter costs. Nevertheless, the figure also indicates that we achieve break-even point after 3 years, and we are actually obtaining interesting profits over a 5-year span.

TABLE 11.5
Costs and Savings Prediction in the Geroskipou Pilot

Concept	Amount (€)	Frequency
Product Costs		
License	16,065	One off payment
Maintenance fee	6,449	Yearly
Customization	0	One off payment
Adoption Costs		
IT Infrastructure	5,000	One off payment
Organizational changes	30,000	One off payment
Training	2,000	One off payment
Others	10,450	One off payment
Savings	27,200	Yearly

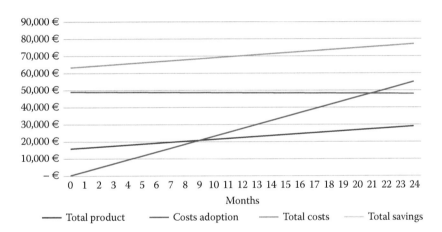

FIGURE 11.5   Costs and savings due to INCA in the Geroskipou Pilot.

11.4.6   CONCLUDING REMARKS ON THE GEROSKIPOU/CYPRUS PILOT

After overcoming initial barriers, the Geroskipou/Cyprus Pilot is now running successfully. The pilot aims at improving the quality of life of the citizens and is not primarily required to generate any income from the service, as political sponsors will assist with financial contributions to ensure the financial sustainability of the pilot. Moreover, Geroskipou plans to implement INCA as a core solution in the new medical center to serve the residents of the whole Paphos District. Hence, overall, the INCA pilot in Geroskipou/Cyprus can be seen as one success story of integrated care put to action.

11.5 THE CROATIA PILOT

In the city of Rijeka, the proportion of the population aged over 60 years is around 27%. With increasing age, the number of the population suffering from mental health disorders will increase to over 50%. In such circumstances, it was realized that the co-ordination of health and social services would be beneficial. In any case, before the introduction of INCA, coordination within healthcare (primary, secondary, and tertiary), as well as between social care and healthcare, was clearly insufficient.

Rijeka has used INCA's Care Manager Interface to create one care program: Health and social care for the patients/care users with mental health diseases or other mental health disorders. The main goal is to improve accessibility of health and social services by using INCA for coordination of the providers of social and health services.

Patient's classification relies on an active collaboration between sanitary and social providers, as patient's classification results from combining information already in hands of medical and social providers. In this implementation, there are significant legal problems when it comes to the interchange of social and medical information. The trusted mediator/contact point for the patient that could legally have access to medical as well as social information is the family doctor. However, till now, whenever the family doctors were supposed to take over tasks that are related with the patients' social well-being, they were reluctant to implement them, as they thought that it is not part of their job. On the other side, owing to legal reasons, the social side cannot be the trusted single contact point, since social workers are not legally qualified to receive medical information, not even with the patient's consent. To overcome this obstacle, the family doctors are now receiving financial incentives to take over social tasks, paid on a per task basis.

Besides these required financial incentives, initial pilot results indicate that the implementation of socio-sanitary care pathways makes economic and qualitative sense. At the end of the pilot, it is likely that the INCA trial will reveal that prevention (e.g., the early detection of dementia) is much more cost-effective than reactive interventions. Once sustainability of socio-sanitary care integration in general is solidly established, and proven in practice, it would get assessed by the key stakeholders in Croatian eHealth ecosystem. Depending on its assessment, the operator of the Croatian eHealth IT system would need to tender the socio-sanitary care integration tool in a public procurement process. In the tendering process, usually, it is the cheapest offer that matches the specified criteria that wins. Public procurement has to be done for purchases above € 25,000. Further, the budget is annual for a single license price, and then, yearly reoccurring maintenance payments that cannot exceed 20% of the procurement price can be charged, without further need for tendering.

Before the tendering process, detailed technical and functional specifications need to be prepared. Those, in general, describe the need of the system's end users but are written in a broader context, giving no favor to any of the existing solutions, piloted or otherwise available on the market. Therefore, although INCA is a pilot, already implemented and running, it needs to go through the public procurement process and win it. Therefore, talking about sustainability is an exercise that we can only envision but not ensure.

11.6   THE LATVIA PILOT

Since 2010, Latvia is running an eHealth program for more efficient use of information and communication technology tools. This eHealth program provides for cooperation in the exchange of data between national-level database and business applications, but it does not provide a platform for cooperation between patients, home care providers, family doctors, and specialists. This is where INCA comes in, providing an integrated virtual platform to engage both.

11.6.1   Barriers Found in the Latvia Pilot

Experience during pilot implementation has shown that there are several barriers that make INCA adoption difficult. They are listed as follows:

- Patients' economical barriers for purchasing IT equipment (patients use cell phones/voice calls instead of computers)
- Low ICT skills of general population aged over 65 years
- Organizational and economical issues in potential customer organization
- Lack of financing and specialists to adopt this technical solution
- Medical institutions use call center options (to attend the voice calls mentioned above) and do not have human resource capacity or finances for extra costs
- Low trust on foreign e-system providers

11.6.2   Sustainability of the Latvian Pilot

The sustainability of the Latvian INCA pilot is ensured after the project, provided the technological provider maintains the software in operation after the project ends, without additional costs, and the Latvia partners have enough resources to keep it alive. Clearly, this excludes issues such as adjustments or hosting, which would produce additional costs to be paid to the technological provider.

11.6.3   Considerations Regarding a Broader Role of INCA in Latvia

Apart from pilot sustainability, there are two important issues to sustainability in Latvia, in general.

There are two national providers in Latvia, that is, *BlueBridge* and *Arstu birojs*, that do not allow interoperation of their systems on stationary and ambulatory care with each other or with third-party providers such as INCA.

However, both providers might be very interested to add INCA to their product portfolio, the reason being that this addition could be the game changer and may push the solution that does not have INCA out of the market.

In conclusion, INCA could try to take the Latvian market by making use of one of the big providers as reseller. As of now, in terms of price, prices are lower that what can be found in other old member countries.

A realistic price would be € 1000 per year per hospital. The reasons for this low price are as follows:

1. We are dealing with extremely tight budgets.
2. The way the Latvian health system is set up, hospitals are interested to provide the service for the patients in person. More visits to the hospital mean more financial compensation for the hospital, which is why INCA, in its present form, cannot provide savings.

The reason for the nonexisting savings can be found in the financing model of the hospital. The hospital is receiving money from the state per intervention. Reducing the amount of intervention with more structured care path ways is not in the interest of the hospital. Hence, as long as the incentive system does not change, there will not be savings produced by INCA. Changes would have to take place on a higher political level, which most likely will take quite some time. The usefulness will be in the better organization of tasks and possibly reduced stress levels for hospital workers.

11.6.4 CONCLUDING REMARKS ON THE LATVIA PILOT

Latvia INCA in its present form is clearly not generating savings, which is mainly because of the nature of the health and social system that incentivizes hospitals for every treatment they provide and not for the treatments that can be avoided through proactive measures.

Nevertheless, Latvian pilot has measured high impact on patient satisfaction with the system and will focus more on continuing to provide measures to improve patients' quality of life.

After the INCA project, Ventspils Municipality—in charge of the Latvian pilot—is committed to sustain project results and to continue to work with developed infrastructure. The municipality will find its own recourses to fund work with the pilot patients to establish sustainability—use of competences, infrastructure, and training.

Furthermore, the created database will be integrated in the upcoming national eHealth system, so that, overall, the Latvian INCA pilot can be regarded as another success story of integrated care delivered from the Cloud.

11.7 CONCLUSIONS

Some of the biggest difficulties health systems are facing today are financial, and despite disparities, all countries are meeting the same problems. Aging societies drive the costs of maintaining health and social systems. In addition, the financial crisis puts pressure on health and social systems' budgets.

Proactive integrated socio-sanitary care is seen as a means to lower costs, while improving service quality and exploiting savings potentials. For all Europe, savings could amount to € 100 billion annually. However, healthcare is provided by a plurality of islands, and despite having strong institutions, good hospitals, and primary care getting better organized, they do not work in a coherent system.

Having studied the market for integrated socio-sanitary care, *Integration of Care* is currently more of a concept than a reality. The integration task is not straightforward, and there are a number of bottlenecks and systemic barriers (mainly legal and organizational) at the interface between social care and healthcare settings, such as

- Fragmentation of care and budgeting
- Lack of coordination (political inflight)
- Collaboration

The virtual integration option provided by INCA can help mask some of these issues for end users. Nevertheless, for smaller commercial players, the inhomogeneity of the different healthcare and social care systems constitutes a particular entry barrier, leading to the initial impression that an EU wide rollout can only be achieved with the support of strong local players that know their health and social system extremely well and can collaborate with the INCA Consortium to adapt the INCA solution to the local clients' needs.

For Spanish and Cypriot partners, the financial sustainability numbers indicate that, in the worst case, INCA's financial sustainability is reached after less than 3 years, providing interesting net savings after that. Hence, the INCA proposal is a good offer in a hard time and can be well positioned within professional bodies' strategies, which will have to move forward with their digitalization plans despite the limitations imposed.

For the other two partners (Croatia and Latvia), the situation is different. Overcoming the legacy of communism and, now, with open markets' economies and democracies converging toward the older and more advanced members in the Union, there is certainly still a way to go.

To summarize, the majority of the INCA pilots (3 out of 5), can have cost savings that will quickly outpace INCA deployment, adoption, and maintenance costs. This, together with the positive feedback received on INCA's benefits and features, makes us confident that INCA's sustainability as a commercial project can be achieved after the end of the EU Pilot phase.

ACKNOWLEDGMENTS

This work has received funding from the EU under the project "*Inclusive Introduction of Integrated Care*" (INCA) with grant agreement number 621006. The INCA project that runs from January 1, 2014, to June 30, 2016, is funded under the 7th call of the ICT Policy Support Program CIP-ICT-PSP-2013, with a total budget of € 5.09 M.

The chapter is, in parts, based on the INCA Project Deliverables "D5.1 Market Overview" and "D5.2 Sustainability Strategies" with friendly permission of the copyright holders.

12 Computerized Systems for Remote Pain Monitoring
A Case Study of Ambulatory Postoperative Patients

Nuno Pombo and Nuno Garcia

CONTENTS

12.1 INTRODUCTION

According to the International Association for the Study of Pain (Loeser and Treede 2008), pain is an unpleasant sensory and emotional experience related to past or potential tissue damage, and patients often may describe their pain in these terms. If it prevails during a relatively short duration, it is known as acute pain, whereas if it persists over a long period of time, it is regarded as chronic pain (Apkarian et al. 2009). Pain is a highly subjective and difficult-to-quantify phenomenon, because it is an individual and personal experience (Giordano et al. 2010), and this subjectivity creates challenges in its description, assessment, and treatment. Instead of being assessed as an isolated value, pain results from multiple aspects (Ong and Seymour 2004; Melzack and Casey 1968; Fernandez and Turk 1992; Holroyd et al. 1996; Kornbluth et al. 2008), such as sensory (e.g., location and intensity), affective (e.g., depression and anxiety), and cognitive (e.g., quality of life). For this reason, patients with pain are called to answer many questionnaires and scores and/or to adopt specific behaviors as a way to treat their pain in all its dimensions such as self-monitoring of pain, adherence to prescribed medications, weight control, and/ or daily exercise. In addition, the effect of inadequate pain relief, besides the ethical concerns, may result in premature discharge from the hospital, postoperative complications, a negative impact on function and quality of life (Ashburn and Staats 1999; Langley et al. 2010; Stewart et al. 2003; Roberto and Reynolds 2002; Dalton et al. 2001), and/or an economic burden (Committee on Advancing Pain Research, Care and Medicine 2011; Cousins et al. 2000; Sheehan et al. 1996; Zimberg 2003), as well as an interference in the quality of life and/or physical and mental disorders, such as distress and anxiety (Apfelbaum et al. 2003; Taylor and Stanbury 2009; Breivik 1995; Berman et al. 2009; Campbell et al. 2003; Morrison et al. 2003; Wall 1979). Many indicators suggest a continued growth in the number of ambulatory surgeries (Berryman 1987), primarily for economic reasons, so as to reduce the costs of inhospital patient accommodations. These surgical procedures usually cause discomfort and moderate to severe pain in a significant percentage of patients (McGrath et al. 2004; Mulcahy et al. 2011). Therefore, pain management is an essential care component in ambulatory surgical centers and hospital wards.

In line with this, clinical decision support systems (CDSSs) based on electronic diaries (EDs) have been increasingly used in recent years, with the goal of providing reliable pain assessments, thus producing higher-quality treatments and outcomes. In fact, the adoption of ED as computerized version of paper pain diaries (PDs) enables patients either to report complaints close in time about the occurrence of pain, called ecological momentary assessment (EMA), or to address retrospective pain that results in pain recall over some period of time. Furthermore, ubiquity and interoperability concepts offer opportunity for remote access, to both healthcare professionals (HCPs) and patients, anywhere and at any time, which raises several challenges to the design, development, and sustainability of CDSS.

Therefore, this chapter focuses on CDSS concepts from the healthcare perspective and presents a case study with its application.

The chapter is organized as follows: Section 12.2 describes CDSS; starting from the motivation for its use, several challenges and opportunities are pinpointed; Sections 12.3 through 12.5 present a case study for remote pain monitoring, including a description of its features and technological concepts; finally, Section 12.6 concludes the chapter and summarizes the main points.

12.2   CLINICAL DECISION SUPPORT SYSTEMS

Clinical decision support has been defined as a system or process that helps professionals make clinical decisions to enhance patient care (Vetter 2015). This encompasses systems as diverse as sophisticated platforms to store and manage medical data; tools to alert HCPs of risk of problematic situations; tools to support HCPs on the decision-making for managing patients with specific disease states; and medication prescription, preventive care, or diagnosis support. These systems offer the potential to improve overall care safety, foster evidence-based practice, enhance HCPs' and patients' education, and optimize cost-effectiveness of care, which may help close the gap between healthcare provider knowledge and performance.

Despite a significant improvement of these systems in the last decade, owing to their design, product development, and increased technological complexity, CDSS are facing some challenging tasks as follows:

- Complexity of the decision support process
- Decision-making based on patient-centered care
- Integration with other systems aiming at a smooth combination of different computer tools, mainly to assure that they should be able to operate without manual entry of data already obtained
- Ability to deal with the heterogeneity of clinical data sources, which may differ in the data models, schemas, naming conventions, and granularity to represent similar data
- Interoperability and ubiquity for a permanent and multidevice data acquisition, data access, and data retrieval
- Need up-to-date and accurate information continuously for an adaptive decision-making
- Evaluation and clinical validation

Furthermore, when CDSS technology is introduced, it is crucial to address all the requirements in order to meet the need of users who should interact with the system, namely the HCPs, the patients, and other relevant stakeholders. To design suitable and accurate system, it is desirable to assure that HCPs contribute with their extensive knowledge base and ability to identify and coherently synthesize all the CDSS requirements. Moreover, the adoption of mobile devices, such as smartphones and hand-held tablets, and the Internet connectivity raised the paradigm of the new care model based more on contacts than on visits (Escarrabill et al. 2011). In fact, the ability to interact with the system anywhere and at any time thoroughly changes the coordinates of time and place and offers invaluable opportunities to these healthcare services delivery. In addition, mobile devices showed significant advances in storage capacity, battery efficiency, portability (Keogh et al. 2010), and the ability to access Internet-based resources (Rosser et al. 2009) that increased its suitability to healthcare systems. Thus, these ubiquitous devices are now part of our modern society and are becoming more important in healthcare-related CDSS (Pombo et al. 2014).

As self-reporting PDs are the most reliable method for pain assessment, a study that presents a comprehensive overview on CDSS, including ED, delivered through mobile devices and combined with a web-based to provide clinical data and clinical guidance, is important and timely.

12.3 CASE STUDY

The case study presented in this chapter was conducted at *Hospital Sousa Martins* in Guarda, Portugal, during a 6-week period through specialty care physician referrals from the ambulatory surgery department. These patients had surgical procedures, from which a certain degree of pain was expected or possible during the initial postoperative days. The participants were enrolled after completing a written consent form. In addition, the protocol of the study was approved by the appropriate ethics committee. Inclusion criteria consisted of the following:

1. Age ranging from 18 to 75 years
2. Status I or II in the Scale of Risk of the American Society of Anesthesiologists
3. Basic computer and mobile phone literacy

Patients were not considered for participation if they had any of the following exclusion criteria:

1. Severe physical or mental impairment that precludes the utilization of the mobile device or the use of the software installed on the device
2. Inability to speak Portuguese fluently
3. Previously received cognitive-behavior therapy
4. Previously used computerized devices for pain monitoring

Thirty-seven individuals were assessed for eligibility, and five were excluded based on the above-mentioned criteria. From the five, two had impairments that precluded the use of the mobile device, one was not fluent in Portuguese, and two refused to participate, claiming of not having available time. As depicted in Figure 12.1, the final sample consisted of 32 participants, randomly assigned to two groups by using a 1:1 ratio. Group I (the Intervention Group), comprising 16 patients, submitted to a treatment condition. Group II (the Control Group), comprising 16 patients, did not submit to a treatment condition and were used as a control group. One participant from Group

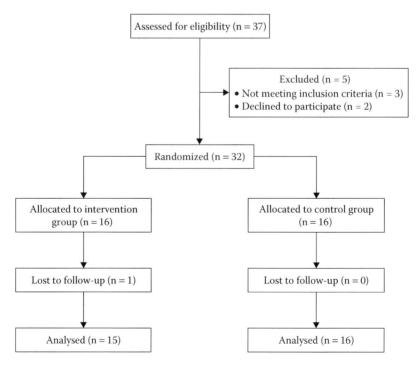

FIGURE 12.1   Flow diagram.

I dropped out during the follow up because of personal reasons. Therefore, participation rate was 86% and our attrition rate was 3.13%. Both treatment groups continued to receive medical care for their pain condition through a specialty medical clinic.

This study encompasses three phases, described as follows:

- *Pretreatment*: Each patient eligible to participate was asked to complete an informed consent agreement and a battery of assessments to obtain baseline values for the outcome measures.
- *Treatment*: All patient-reported outcome measures were obtained by asking the participant to complete a seven-point Likert scale questionnaire during the hospital stay after surgical intervention, supervised by an HCP. The participants in both groups of the study were called by the HCP after 24 h and again after 5 days and were asked to rate their average pain. During the phone interviews, data were entered directly into the monitoring software. Study personnel assigned to assist participants in the clinic setting were informed of participants' treatment assignment. In addition, a daily electronic PD was used to assess self-reported pain by the participants of the Intervention Group during the 5-day computerized monitoring program period. Participants were asked to complete several pain ratings per day, using an 11-point numerical rating scale (NRS) with anchors at $0 =$ no pain to $10 =$ worst pain, commonly in morning, afternoon, and evening, in accordance with the treatment protocol. Each participant was registered on a web-based personal health record (PHR) and a mobile application (app) corresponding to an electronic pain diary (Pombo et al. 2012), which was installed on the smartphones dispensed to every participant of the Intervention Group.
- *Posttreatment*: As part of the computerized monitoring program, participants in the Intervention Group completed an additional questionnaire to evaluate their adherence and experience with the technology, as applied to postoperative home-based pain monitoring. Moreover, participants in the wait-list control group continued with the medical care recommended by their physicians, which for all patients involved a 1-month posttreatment visit at the hospital.

The HCPs access the PHR online to define the patient-oriented treatment (in terms of duration), pain record density, medication frequency, rules, and subsequent content of auto-generated messages, according to the collected values and the patient symptoms. These rules are composed of a series of IF THEN rules and may differ not only among patients who belong to the same intervention but also in accordance with monitoring purposes, participant symptoms, and duration of the intervention. Furthermore, each participant is provided with an Android smartphone that includes an app that periodically checks for changes in treatment configuration; thus, it is always current, according to the clinical settings planned by the HCPs. It is expected that treatment adjustments during the monitoring period will be more responsive to patient needs because clinical visits are not required. The PHR allows the HCPs to consult the data obtained for each patient, supported by a histogram composed of the pain records. The app remains in the background until the scheduled time to take medication and/or register pain approaches. In either case, the patient is alerted with an audible alarm. Medication information comprises textual information, whereas the request for the instantaneous value for pain presents an NRS to the patient for a set period of time. When this time is exceeded, a "no response," indicated by a null value, is registered. All values registered are saved locally, using an SQLite database, and sent to the remote PHR via WS (developed in Windows Communication Foundation), immediately after recording. Then, the app returns to the background. When communication fails during the send process, the value is marked as pending and it is included in the next data transmission. This process is automatic and does not require patient intervention. In addition, the app allows the patient to register unplanned pain records with an identical submission process at times other than the scheduled times. Whenever a message is received, it is saved in the SQLite database, the app is activated, and the text is presented to the patient. The app activation occurs only during the patient's waking time. The collected data are accessible for consultation in the PHR through patient identification, composed of a username and password.

12.4   RESULTS

The final sample consisted of 31 patients (14 males and 17 females) aged between 20 years and 72 years. Intervention Group was composed of 15 Caucasian patients (7 females and 8 males) aged 48.07 ± 12.23 years (mean $\pm$ SD). Control Group was composed of 16 Caucasian patients (10 females and 6 male) aged 50.13 ± 10.79 years (mean $\pm$ SD). Participants were referred to the treatment study for hand pain (48.4%), pelvic pain (38.7%), knee pain (9.7%), and leg pain (3.2%). Treatment groups were equivalent in age, gender, and race ($p > .05$). There were some differences in terms of pain location from the mixture of acute postoperative pain conditions.

The pretreatment questionnaire aimed at characterizing the participants in terms of mobile phone and health services experience (see Table 12.1). Participants in both arms of the study regularly used mobile phones (100/93.3%) (Question, Intervention Group/Control Group %). In contrast, a reduced use of mobile phones was reported for leisure (13.3/6.7%), professional purposes (13.3/0%), and Internet access (13.3/6.7%),

---

TABLE 12.1
Pretreatment Questionnaire

Questions

Q.1	Do you use a mobile phone regularly?
Q.2	Do you use a mobile phone to make/receive calls?
Q.3	Do you use a mobile phone for leisure and/or to play games?
Q.4	Do you use a mobile phone to run software specific to your professional activities?
Q.5	Do you use a mobile phone to access the Internet?
Q.6	Do you know about electronic health records such as Meu Sapo Saúde and Plataforma de Dados de Saúde?
Q.7	Do you subscribe to an electronic health record?
Q.8	Do you use your electronic health record regularly?
Q.9	Do you keep your electronic health record current?
Q.10	Do you consider the use of an electronic health record beneficial?

TABLE 12.2

Posttreatment Questionnaire

Questions

Q.1	Do you consider the application easy to use?
Q.2	Do you consider the training provided by the HCP to be suitable?
Q.3	Do you consider the application to be attractively designed?
Q.4	Do you consider that the terminology is clear and understandable?
Q.5	Do you consider the font color and size easy to read on the screen?
Q.6	Do you consider the response time of the application fast enough?
Q.7	Do you consider the alarm sound easily audible?
Q.8	Do you consider the application suitable for access to the medical indications?
Q.9	Do you consider the application suitable to improve the management of postoperative pain?
Q.10	Do you recommend the application?

the combination of which is unsurprisingly greatly correlated $\left((r_s = .877, p < .01)\right)$. In addition, despite the sense of the benefits that may result from the use of the PHR (46.6/60%), its knowledge, its use, and registration for it (13.3/6.7%) remain almost nonexistent, independent of the patient's age, pain conditions, or symptoms.

The analyses of the posttreatment questionnaire (see Table 12.2) reveal a very strong correlation $(r_s = .844, p < .01)$ between adequate training provided by the HCPs and the ease of use of the application. Moreover, adequate training provided by the HCPs is strongly correlated with the suitability of the application to improve pain management $(r_s = .675, p < .01)$, with the recommendation of the application $(r_s = .750, p < .01)$ and with the clarity and understanding of the terminology used in the application $(r_s = .626, p < .05)$. In addition, design and performance are strongly correlated $(r_s = .843, p < .01)$. The audibility of the alarm sound is strongly correlated with the suitability of the application, both to provide medical information $(r_s = .667, p < .01)$ and to improve pain management $(r_s = .695, p < .01)$, together with the recommendation of the application $(r_s = .666, p < .01)$. In addition, the audibility of the alarm sound is strongly correlated with the suitability of the application to improve pain management $(r_s = .688, p < .01)$ and very strongly correlated with design $(r_s = .751, p < .01)$ and terminology concepts $(r_s = .857, p < .01)$.

The analysis of both pretreatment and the posttreatment questionnaires reveals a strong correlation between the ability to use the mobile phone to make and receive calls and the suitability of the application to provide medical information $(r_s = .704, p < .01)$, together with the positive effects on health from participation in the study $(r_s = .516, p < .05)$.

12.5   DISCUSSION

This study proved that the computerized system, which combines a web-based PHR and mobile devices, is feasible, patients are compliant, and they consider the device user-friendly. This extends previous work on pain monitoring [21,25,35–40], demonstrating its acceptability, patient satisfaction, and patient compliance with computerized treatment in patients with mixed acute pain conditions. Specifically, with respect to the device used in this study, a majority of participants recommends the system and recognizes that it is appropriate for pain management; it is user-friendly and does not require advanced skills or experienced users. These findings are even more significant, because the participants were primarily middle-aged and presented a high degree of illiteracy in terms of handling applications on mobile devices and/or Internet access. The current study provides the evaluation of a purely mobile- and web-based, no-contact intervention for use in the context of routine care. Such a no-contact intervention has the advantage of being broadly available, which may be critical for providing access to a large number of patients. The pain-monitoring system could have major implications if accessed more widely, enhancing the potential societal benefits in terms of pain management and well-being (Djulbegovic et al. 2011).

Furthermore, the inclusion of the PHR in the monitoring system enabled the reliable message delivery required for emergency messages in a fully automated fashion, as well as scalability to support as many patients as possible, with online persistent data available to both patient and HCPs. The PHR was suitable for pain monitoring, providing ubiquitous and real-time access and allowing an effortless definition and management of patient-oriented treatment rules, with minimal therapy. The guidance of the HCPs at the onset of the monitoring is crucial to patients' satisfaction with the system, as evidenced by the high correlation between the recommendation of the application and its suitability to improve pain management, and to provide medical information. The absence of detected and reported errors related either to the app or to the PHR suggests that the proposed system is stable and reliable. Because the electronic PD is based on periodic alarms in accordance with medical protocol, the audibility of the alarm sound is crucial to the system adherence and accuracy.

12.6 CONCLUSION

This chapter highlighted the importance of CDSS for remote pain monitoring, starting from the collected data in the context of the ambient assisted living and its capability to produce accurate and reliable outcomes for healthcare assistance professionals in the clinical decision-making. In line with this, an overview related to mobile devices and web-based remote monitoring focused on complementary concepts such as ubiquity, interoperability, and data acquisition, transmission, and analysis. Finally, a case study was presented related with patients with ambulatory surgery for a 5-day computerized monitoring program period. The main purpose was to evaluate a combined web-based and mobile computerized system for short-term pain monitoring. The main conclusions are as follows:

- The study proved that the proposed computerized system is feasible.
- The study proved the feasibility of a purely mobile and web-based, no-contact intervention for use in the context of routine care. Such a no-contact intervention has the advantage of being broadly available, which may be critical for providing access to a large number of patients.
- The proposed system is user-friendly and does not require advanced skills or experienced users.

Further studies should be addressed to determine the economic effects of the proposed monitoring model not only for patients but also on the healthcare system. Finally, further work is needed to evaluate the proposed system to follow up on participants for longer periods of time, for example, encompassing patients with chronic pain symptoms.

ACKNOWLEDGMENTS

This work was supported by Fundação para a Ciência e a Tecnologia (FCT) project **UID/EEA/50008/2013** (Este trabalho foi suportado pelo projecto FCT UID/EEA/50008/2013). The authors acknowledge the contribution of COST Action IC1303 - AAPELE - Algorithms, Architectures, and Platforms for Enhanced Living Environments. The authors also acknowledge the contributions of the staff at the Ambulatory Surgery Department of *Hospital Sousa Martins*, Guarda, Portugal.

REFERENCES

Apfelbaum, Jeffrey L., Connie Chen, Shilpa S. Mehta, and Tong J. Gan. 2003. Postoperative Pain Experience: Results from a National Survey Suggest Postoperative Pain Continues to Be Undermanaged. *Anesthesia & Analgesia* 97(2):534–540. doi:10.1213/01. ANE.0000068822.10113.9E.

Apkarian, A. Vania, Marwan N. Baliki, and Paul Y. Geha. 2009. Towards a Theory of Chronic Pain. *Progress in Neurobiology* 87(2):81–97. doi:10.1016/j.pneurobio.2008.09.018.

Ashburn, Michael A. and Peter S. Staats. 1999. Management of Chronic Pain. *The Lancet* 353(9167):1865–1869. doi:10.1016/S0140-6736(99)04088-X.

Berman, Rebecca L. H., Madelyn A. Iris, Rita Bode, and Carol Drengenberg. 2009. The Effectiveness of an Online Mind-Body Intervention for Older Adults with Chronic Pain. *The Journal of Pain* 10(1):68–79. doi:10.1016/j.jpain.2008.07.006.

Berryman, J. M. 1987. Development and Organization of Outpatient Surgery Units: The Hospital's Perspective. *The Urologic Clinics of North America* 14(1):1–9.

Breivik, Harald. 1995. 1 Benefits, Risks and Economics of Post-Operative Pain Management Programmes. *Baillière's Clinical Anaesthesiology* 9(3):403–422. doi:10.1016/S0950-3501(95)80014-X.

Campbell, Lisa C, Daniel J. Clauw, and Francis J. Keefe. 2003. Persistent Pain and Depression: A Biopsychosocial Perspective. *Biological Psychiatry* 54(3):399–409.

Committee on Advancing Pain Research, Care, and Education; Institute of Medicine. 2011. *Relieving Pain in America: A Blueprint for Transforming Prevention, Care, Education, and Research*. The National Academies Press. http://www.nap.edu/openbook. php?record_id=13172 (accessed August 19, 2016).

Cousins, Michael J., Ian Power, and Graham Smith. 2000. 1996 Labat Lecture: Pain–a Persistent Problem. *Regional Anesthesia and Pain Medicine* 25(1):6–21.

Dalton, Jo Ann, John Carlson, William Blau, Celeste Lindley, Susan M. Greer, and Richard Youngblood. 2001. Documentation of Pain Assessment and Treatment: How Are We Doing? *Pain Management Nursing: Official Journal of the American Society of Pain Management Nurses* 2(2):54–64.

Djulbegovic, Benjamin, Ashly D. Black, Josip Car, Claudia Pagliari, Chantelle Anandan, Kathrin Cresswell, Tomislav Bokun, et al. 2011. The Impact of eHealth on the Quality and Safety of Health Care: A Systematic Overview. *PLOS Medicine* 8(1). Article no. e1000387.

Escarrabill, Joan., Tino Marti, and Elena Torrente. 2011. Good Morning, Doctor Google. *Revista Portuguesa de Pneumologia* 17(4):177–181.

Fernandez, Ephrem, and Dennis C. Turk. 1992. Sensory and Affective Components of Pain: Separation and Synthesis. *Psychological Bulletin* 112(2):205–217.

Giordano, James, Kim Abramson, and Mark V. Boswell. 2010. Pain Assessment: Subjectivity, Objectivity, and the Use of Neurotechnology. *Pain Physician* 13(4):305–315.

Holroyd, Kenneth A., France Talbot, Jeffrey E. Holm, Jeffrey D. Pingel, Alvin E. Lake, and Joel R. Saper. 1996. Assessing the Dimensions of Pain: A Multitrait-Multimethod Evaluation of Seven Measures. *Pain* 67(2–3):259–265.

Keogh, Edmund, Benjamin A. Rosser, and Christopher Eccleston. 2010. E-Health and Chronic Pain Management: Current Status and Developments. *Pain* 151(1):18–21. doi:10.1016/j.pain.2010.07.014.

Kornbluth, Ira D., Mitchell K. Freedman, Michele Y. Holding, E. Anthony Overton, and Michael F. Saulino. 2008. Interventions in Chronic Pain Management. 4. Monitoring Progress and Compliance in Chronic Pain Management. *Archives of Physical Medicine and Rehabilitation* 89(3):S51–S55.

Langley, Paul, Gerhard Muller-Schwefe, Andrew Nicolaou, Hiltrud Liedgens, Joseph Pergolizzi, and Giustino Varrassi. 2010. The Impact of Pain on Labor Force Participation, Absenteeism and Presenteeism in the European Union. *Journal of Medical Economics* 13(4):662–672. doi:10.3111/13696998.2010.529379.

Loeser, John D. and Rolf-Detlef Treede. 2008. The Kyoto Protocol of IASP Basic Pain Terminology. *Pain* 137(3):473–477. doi:10.1016/j.pain.2008.04.025.

McGrath, Brid, Hany Elgendy, Frances Chung, Damon Kamming, Bruna Curti, and Shirley King. 2004. Thirty Percent of Patients Have Moderate to Severe Pain 24 Hr after Ambulatory Surgery: A Survey of 5,703 Patients. *Canadian Journal of Anesthesia* 51(9):886–891. doi:10.1007/BF03018885.

Melzack, Ronald, and Kenneth L. Casey. 1968. Sensory, Motivational, and Central Control Determinants of Pain: A New Conceptual Model. *The Skin Senses*, 423–443.

Morrison, R. Sean, Jay Magaziner, Mary Ann McLaughlin, Gretchen Orosz, Stacey B. Silberzweig, Kenneth J. Koval, and Albert L. Siu. 2003. The Impact of Post-Operative Pain on Outcomes Following Hip Fracture. *Pain* 103(3):303–311.

Mulcahy, James J., Shihong Huang, Junwei Cao, and Fan Zhang. 2011. How Are You Feeling? A Social Network Model to Monitor the Health of Post-Operative Patients. In *Systems Conference (SysCon), 2011 IEEE International*, pp. 149–154. doi:10.1109/SYSCON.2011.5929127.

Ong, K. S., and R. A. Seymour. 2004. Pain Measurement in Humans. *The Surgeon* 2(1):15–27.

Pombo, Nuno, Pedro Araújo, and Joaquim Viana. 2014. Knowledge Discovery in Clinical Decision Support Systems for Pain Management: A Systematic Review. *Artificial Intelligence in Medicine* 60(1):1–11. doi:10.1016/j.artmed.2013.11.005.

Pombo, Nuno, Pedro Araújo, Joaquim Viana, Benjamin Junior, and Rita Serrano. 2012. Contribution of Web Services to Improve Pain Diaries Experience. In *Lecture Notes in Engineering and Computer Science: Proceedings of the International MultiConference of Engineers and Computer Scientists IMECS 2012*, Vol. 1, March 14–16, Hong Kong, pp. 589–592. http://www.iaeng.org/publication/IMECS2012/IMECS2012_pp589-592.pdf (accessed August 19, 2016).

Roberto, Karen A., and Sandra G. Reynolds. 2002. Older Women's Experiences with Chronic Pain: Daily Challenges and Self-Care Practices. *Journal of Women & Aging* 14(3–4):5–23. doi:10.1300/J074v14n03_02.

Rosser, Benjamin A., Kevin E. Vowles, Edmund Keogh, Christopher Eccleston, and Gail A. Mountain. 2009. Technologically-Assisted Behaviour Change: A Systematic Review of Studies of Novel Technologies for the Management of Chronic Illness. *Journal of Telemedicine and Telecare* 15(7):327–338. doi:10.1258/jtt.2009.090116.

Sheehan, J., J. McKay, M. Ryan, N. Walsh, and D. O'Keeffe. 1996. What Cost Chronic Pain? *Irish Medical Journal* 89(6):218–219.

Stewart, Walter F., Judith A. Ricci, Elsbeth Chee, Steven R. Hahn, and David Morganstein. 2003. Cost of Lost Productive Work Time among US Workers with Depression. *JAMA* 289(23):3135–3144.

Taylor, Allison, and Linda Stanbury. 2009. A Review of Postoperative Pain Management and the Challenges. *Current Anaesthesia & Critical Care* 20(4):188–194. doi:10.1016/j.cacc.2009.02.003.

Vetter, Mary Jo. 2015. The Influence of Clinical Decision Support on Diagnostic Accuracy in Nurse Practitioners. *Worldviews on Evidence-Based Nursing* 12(6):355–363. doi:10.1111/wvn.12121.

Wall, Patrick D. 1979. On the Relation of Injury to Pain the John J. Bonica Lecture. *Pain* 6(3):253–264. doi:10.1016/0304-3959(79)90047-2.

Zimberg, S. E. 2003. Reducing Pain and Costs with Innovative Postoperative Pain Management. *Managed Care Quarterly* 11(1):34–36.

13 Simple Rehabilitation Games for Special User Groups

Antti Koivisto

CONTENTS

13.1 INTRODUCTION

It is important to maintain people's well-being throughout their lives. In order to achieve this goal, people should have a possibility of exercising physically and mentally. People with special needs and the elderly people are too often considered a minor and marginalized group, who have no use or even interest in game technology. Our goal was to create easily approachable games with low costs that suit the majority of people, despite physical or cognitive limitations. We used mobile phones as game controllers, as the majority already own a mobile phone and it is easy to get them for a reasonable prize. Mobile phones also have several built-in sensors that can be used to track almost any kind of movement with the right algorithms. In addition, they have other features that can be used to enhance well-being in terms of games. These types of serious games can be called exergames or rehabilitation games, depending on the purpose of the game.

Several studies have shown that both physical exercise and game playing have positive effects on people, including older adults, people with learning disabilities, people combating serious depression, and even people suffering from Alzheimer's disease (e.g., Fairchild & Scogin, 2010; Geda et al., 2010; Spector et al., 2003; McCough et al., 2011). Findings from scientific research show that, in general, playing video games can lead to changes in an individual's pleasure, vigilance, and dominance and, therefore, in their overall state of experienced well-being. In addition, in the case of older adults, simple and easy-to-play video games are well accepted and found to create positive feelings and enjoyment (Khoo & Cheok, 2006; Koivisto et al., 2013; Sirkka, Merilampi, Koivisto, Leinonen, & Leino, 2012; Snowden et al., 2011). Even a few minutes of regular game exercise daily will produce cognitive benefits and improve performance requiring skills, for example, attention and concentration (Gao & Mandryk, 2012b).

Our research group respects the end user highly. Therefore, user experiences and usability are the most important subject matters in our research. We have conducted several interviews to explore how various target groups adopt our games and how they feel while playing the games. By these studies, we have tried to find out how to trigger the players' flow state, because flow seems to have a positive influence on performance enhancement, learning, and engagement, which are all important goals in serious games (e.g., Csikszentmihalyi, Abuhamdeh, & Nakamura, 2005; Engeser & Rheinberg, 2008). As a result, we have to ensure that there is an appropriate balance between personal skills and the game play. Moreover, when making exergames for our special user groups, we have to produce a proper balance between the game intensity and the players' fitness level. On top of all this, the games should be suitable for the majority of people, despite their physical or cognitive limitations.

In this chapter, we present three different cases of cognitively stimulating mobile games that include light cognitive and/or physical exercises. We also present the technology behind those games. The games are tested with different target groups: older adults, people with learning disabilities, and people with physical disabilities. This chapter also discusses the subjective experiences of the participants and staff observations related to these trials.

13.2 GAME DESIGN

The three different game cases that are presented in this study are categorized on the basis of the game type and gaming control. The first game (Case #1) is a game that is controlled by tilting the mobile phone placed on a balance board. The game is played on a large screen. Case #1 game consists of mild physical exercise. The second game (Case #2) is played with a tablet PC. This game is controlled by tilting (cat vs. mouse game) or by tapping (trail-making game) the tablet. Case #2 game consists of gaming and cognitive exercise. The third game (Case #3, balance board painting) is controlled by a mobile phone and a tablet PC. The mobile phone is placed on a balance board and the tablet is held in hand. The game can be steered by hands or by feet with the whole body by standing on top of the balancing board. Case #3 game, this is, the balance board painting, combines game technology, with fine arts as a means for physical rehabilitation.

Case #1 game was tested in trials with older adults and people with learning disabilities. A cognitively stimulating game in Case #2 was tested with a group of older men diagnosed with memory impairment. Case #3 game (balance board painting) was tested in collaboration with the rehabilitation center of Kankaanpää and a nursing home for disabled people called Kaunummen Koti. We also carried out a test during a regional senior fair. The test participants in Case #3 were mainly older adults or people with physical disabilities.

We paid special attention on the game design owing to our user group with special needs. Because of the impaired perception and sensation skills in the target groups, the following accessibility principles were deployed in the game design: large target button elements, simplified and only necessary graphics, minimal amount of animation, colors used conservatively with high contrast, simple one-view display at a time, and important information in the middle of the screen (Diaz-Bossini & Moreno, 2014). In addition to the above-mentioned accessibility principles, the games were designed to use obvious logic without additional introductions on how to play. This was considered especially important for people with learning disabilities, even when assisted by the staff to start the game play (Sirkka et al., 2014).

A time-up limit was also added to all three games. This limit was needed to guarantee the achievement of both goals in game play, that is, being entertaining and rehabilitating. Multiple but short gaming sessions with moderate intensity on a regular basis have been proved to be more beneficial than less frequent but longer play sessions (Gao & Mandryk, 2012a, 2012b).

In general, the games should correlate with the physical condition of the player as well as with their skills to achieve the so-called "flow feeling." The games are designed to generate a positive effect on players and are most successful and engaging when they facilitate the flow experience. A flow state can be entered while performing any activity, although it is most likely to occur when one is performing a task or activity for intrinsic purposes wholeheartedly (Csikszentmihalyi, 1975). Passive activities such as taking a bath or even watching TV do not usually elicit flow experiences, as individuals have to do something actively to enter a flow state.

13.2.1 Flow Theory

Csikszentmihalyi (1975) introduced the concept of flow state when studying people involved in activities such as rock climbing, chess, and dance. Subsequently, the flow theory has been applied in several different domains, including sports, art, work, human–computer interaction, games, and education (Kiili, Freitas, Arnab, & Lainema, 2012). Flow refers to a state of complete absorption or engagement in a specific activity in which a person excludes all irrelevant emotions and thoughts (Csikszentmihalyi, 1990). During an optimal experience, a person is in a positive psychological state, where he or she is so involved with the goal-driven activity that nothing else seems to matter. An activity that produces such experiences is so pleasant that the person may be willing to do something for its own sake, without being concerned with what he or she will get out of this action. This kind of intrinsic motivation is very important, especially in serious games that usually require different cognitive or physical investments than entertainment games.

Csikszentmihalyi (1990) has distinguished nine flow dimensions that constitute the flow experience. These dimensions can be further divided into flow conditions and flow characteristics (Swann et al., 2012). Flow conditions are prerequisites of flow, and they are also called antecedents of flow. The flow theory postulates that there are three conditions that have to be fulfilled to achieve a flow state:

1. The person is involved in an activity with a clear set of goals and progress.
2. The task at hand provides clear and immediate feedback.
3. There must be a balance between the perceived challenges of the task at hand and the person's own perceived skills.

The flow characteristics describe the feelings of an individual when experiencing flow. Flow characteristics include sense of control, action-awareness merging, loss of self-consciousness, concentration, time distortion, and autotelic experience dimensions. On the other hand, it has been argued that the combination of the first eight dimensions of flow leads to flow that is characterized as an autotelic experience (9th dimension). Autotelic experience refers to enjoyable and intrinsically rewarding experiences. The literature shows that flow researches have not reached a definitional agreement on the division of the dimensions into conditions and characteristics. Despite the fact that Csikszentmihalyi (1990) has argued that whenever people reflect on their flow experiences, they mention some, and often all nine flow dimensions.

Theoretically, flow consists of nine dimensions, but in particular, immediate feedback, sense of control, loss of self-consciousness, clear goals, and the challenge-skill balance dimension provide a meaningful approach, with which to embody engaging elements into exergames used in activation (Figure 13.1) (Kiili, 2005; Kiili et al., 2012, 2013; Koivisto et al., 2013; Merilampi, Sirkka, Leino, Koivisto, & Finn, 2014; Wilson et al., 2002).

13.2.2 GAME DESIGN SUMMARY

In conclusion, our game designing focused on three factors:

1. Set focus on accessibility factors in the games
2. Use minimal amount of equipment in order to play the games
3. Make the user achieve "the flow feeling" when playing, to make game play activating

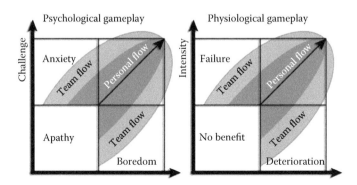

FIGURE 13.1 Dual-flow model.

The third part was achieved by making the games appealing, easy, clear, and fun to play. However, the game should be a "serious game," with a specific goal and purpose applicable in rehabilitation. These requirements led us to select simple and easy-to-use components such as a mobile phone, an Internet connection, a tablet PC, and a TV display. These gaming components could be hidden easily enough for technology, not to become a barrier to our user groups.

13.3 OVERVIEW OF THE GAMES AND TECHNOLOGY INSIGHT

Because of our user groups, we did not want to create visually too fully stowed graphical triumph. Instead, we decided to make the games as simple as possible and hide all technical devices from the user groups, if possible. This section describes the game ideas in each test case, the technology used in the game, and the reason for why we selected the game for the user group.

13.3.1 CASE #1: BALANCE BOARD GAMING

In Case #1, balance board gaming, the game included mild physical exercise. Mild exercise was chosen because of our test groups, who may not be self-motivated to exercise. The game was tested in two different places, with different user groups. The first testing occasion was carried out with older adults in nonrecurring game events, where volunteers had an opportunity to play the game as many times as they wished. Professional staff, researchers, and other game testers observed the game play. The second test occasion took place with people with learning disabilities, all of whom worked in a sheltered work facility.

The game concept in balance board gaming employed a mobile phone embedded with an acceleration sensor (1) to control the game play on a TV screen. By constantly reading the acceleration data, the phone communicates it to a socket server via the Internet (2). The socket server then determines the moves of the game characters and displays them on the screen (3). Figure 13.2 shows how the data are handled and transferred in the system between the different devices.

The reason for selecting a socket server was to provide the communication framework with a multiplayer possibility in the trials. We selected Node.js as a socket server, because we gained the best possible interaction between the HTML5 game (on a browser) and mobile phones (game controllers) with Node.js. Node.js is an open-source, cross-platform runtime environment for the server-side and networking applications. A server Socket. IO was installed in the Node.js. It is a JavaScript library for real-time web applications. IO enables real-time, bidirectional communication between the web clients and the server.

We developed an application for the mobile phone by using a PhoneGap framework. PhoneGap is a free and open-source framework that enables the use of JavaScript, Hyper Text Markup Language (HTML), and cascading style sheets (CSS), instead of relying on platform-specific application programming interfaces (APIs), for example, those used in iOS or Android. In addition, PhoneGap allows plugins to access the phone's hardware quite easily. This is usually needed in order to get the most of the application. In this case, the game controller could have been completed without a mobile application by running a web page in a mobile phone.

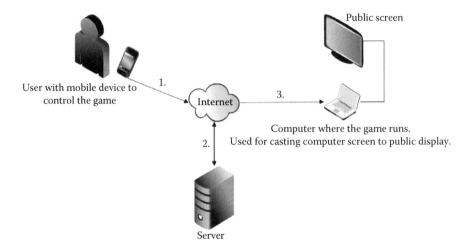

FIGURE 13.2   Communication between devices and the gaming server.

However, by creating an application, we simplified the whole playing situation. Moreover, data transmission was possible even when the screen was turned off. By turning the screen off, misclicks were prevented. If the game control had been a web page, the data transmission would have been stopped when the phone went to sleep. This in turn would have terminated the play.

The game was stored on the web for anyone and everyone to join the gaming session, with the right app on their phones. New TVs even provide a possibility of viewing web pages and thereby make an external gaming console unnecessary. The only equipment needed for game play was a mobile phone, a TV, and an Internet connection. The game itself was made with JavaScript by using an HTML canvas element to draw the game events on.

The game focused on practicing the motor coordination of hands or feet by catching the red square/cheese with a purple ball/mouse. In addition, the game provided cognitive stimulation, as the player must react/observe where the next square/cheese appears (Sirkka et al., 2012). The design was kept visually clear and logically simple. The game field has impenetrable obstacles, which have to be bypassed (Figure 13.3).

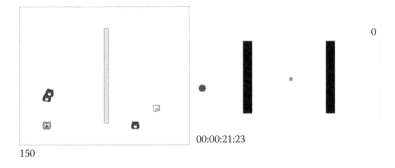

FIGURE 13.3   Case #1, balance board gaming: screenshot of the game.

The game was made slightly different for the older adults and people with learning disabilities, in order to maintain their interest but still keep the game as simple as possible. This was completed by changing the game elements. Otherwise, the main difference was that cats appeared in the game designed for people with learning disabilities. The game starts with no opponents, except the impenetrable obstacle to be bypassed. The game becomes more challenging after collecting five cheese chunks. After that, the first opponent, a cat, joins the game and starts chasing the mouse. Another cat appears after 10 cheese chunks have been collected, while the first cat gets faster and harder to evade. The third cat appears after 15 cheese chunks have been collected, and the two existing cats start chasing the mouse even faster. The game terminates when the 1-minute time limit is over or when a cat catches the mouse.

The movement of the ball/mouse is controlled by tilting the mobile phone. The steeper the tilt, the faster the ball/mouse moves on the screen. Two different controlling modes were used: (1) manually by wrist movement or (2) foot-controlled movement, with the phone attached to a balance board. In both modes, the players were seated during the game for safety reasons. The idea of the game is to collect as many red squares/cheese chunks as possible in 1 minute. The physical challenge can be increased if the game is played by standing. In that case, the whole body needs to be moved to control the game.

13.3.2 Case #2: Tablet PC Gaming

The games in Case #2 focused on gaming and cognitive exercise. In the tablet PC gaming case, two different games were developed in collaboration with memory rehabilitation professionals:

1. Cat vs. mouse game, which is similar to the game presented in Case #1, balance board gaming
2. Trail-making test game, which is an electronic version of a traditional memory test

The games were made to stimulate memory, coordination, reaction, and attention skills. In this case, the user group consisted of older adults diagnosed with memory impairment. The games were controlled by tilting the tablet PC or by tapping the touch screen.

The game design of the cat vs. mouse game was modified into a tablet PC version, which simplified the gaming event and data communication between the tablet and the server. Otherwise, the game play was identical to the game in Case #1 (balance board gaming). As seen in Figure 13.3, the only difference between these games was that the phone was replaced by a tablet as the game controller, thus eliminating additional computer-Internet-TV screen requirements. The feedback of previous trials with Case #1 (balance board gaming) showed that animal characters helped the players catch the logic of the game. They also made the game more exiting with gradually increasing levels of difficulty. This is why, it was used in the testing events.

The trail-making test game is a modified version of the Trail Making Test A (TMT-A) used in memory diagnostics. The standardized test distributes the numbers 1–25 randomly over a sheet of paper. The tester should draw a line, connecting the numbers in ascending order as quickly as possible, without lifting the pen from the paper. The time of drawing the trail is measured. The tester is allowed to correct possible errors, but it increases the completion time reported in seconds. The longer the total time, the greater the memory impairment (Alaska Department of Administration, 2013; Tombaugh, 2004).

Blue digits are randomly distributed over the screen in the game, and the user is instructed to tap them in ascending order as quickly as possible (Figure 13.4). A correctly tapped digit turns translucent, whereas an error turns red. The size and coloring of numbers were made as visible as possible. The game ends when all digits have been tapped in a correct order after four errors or when the time runs out (240 s). In our previous trials, the difficulty level adjustment was found important, because the testers were different. A menu was built to select a desired difficulty level from 5 to 40 digits in five-digit intervals.

A game score registry system was built to monitor personal progress in both games. Each participant was provided with a personal near field communication (NFC) identification card to log in to the game. When starting the game, the player touched the NFC reader of the tablet with his or her personal card. The game presented the latest records on the screen, based on identification, and then presented correct difficulty levels to the games. The NFC technology registered the data of each play event, enabling the analysis and follow-up of the progress of each individual player. The games were played in a 3-month test period, during which the players played the games for 10 minutes per day. Parallel trials will be carried out in Ireland.

13.3.3 Case #3: Balance Board Painting

The idea in Case #3 (balance board painting) was to combine art and technology to enhance well-being and physical rehabilitation. The idea of mobile painting and that of our prototype was tested on three different occasions (in a rehabilitation center of Kankaanpää, in a nursing home for the disabled, and in a summer event and regional senior fair). The first and the most important target group comprised physically

150

FIGURE 13.4 Case #2, tablet PC gaming: both games.

FIGURE 13.5 Case #3, balance board painting: gaming event.

disabled people. The game applications were tested in collaboration with the reha-bilitation center of Kankaanpää and the nursing home for the disabled, Kaunummen Koti. The testers had different kinds of mobility state, with mild to very challenging constraints. Several workshops were arranged, where the rehabilitees could practice the game. A mobile phone or a tablet PC was used as a game controller. They could be steered either by holding the device in hands or by feet, using a balance board (Figure 13.5). The game was mainly developed for people with serious physical limi-tations, but our test groups varied a lot, since the public had a possibility of testing balance board painting in various expositions in 2014.

The technology used in this case was very similar to the technology used in Case #1, balance board gaming. The users steered a red ball on the screen. The red ball was used to represent a pen tip, and the drawing area was visible all the time. The main difference from the previous game was that the users could use two devices at the same time:

1. The mobile phone was used to determine the position of the pen tip on the screen in the balancing board.
2. The tablet PC in the user's hand was used to determine the color and pen size and for on/off function.

The mobile balance board painting also included the possibility of printing the work of art. This was quite a popular feature in the application, as most of the users wanted to show what they had done while exercising with the mobile painting equipment.

13.4 RESULTS OF THE TRIALS

The results discussed in this chapter have been collected from several trials with various target groups. The games in Case #1 (balance board gaming) and in Case #3 (balance board painting) were tested in nonrecurring game events, where volunteers had an opportunity to play the game as many times as they wished. Professional staff, researchers, and other game testers observed the game play. The subjective experiences and perceptions, including common free comments on the trials, were collected immediately after each player's experiment by a structured interview. The data were analyzed with content analysis by calculating frequencies and percentages in the responses to the themes.

The game in Case #2 (tablet PC gaming) was tested in a 3-month trial setting with male war veterans. The aim of the trial was to measure possible effects of cognitively stimulating mobile games on older people's cognitive skills, when they played on a regular basis. A group of participants in an assisted living environment volunteered for this trial. The user experiences and usability issues were collected by structured interviews immediately after the 3-month gaming period. Each participant was interviewed privately in order to obtain as authentic and subjective information as possible, with regard to their previous experiences and use of mobile games and devices. We collected their subjective experiences of the trial period, its impacts on their cognitive skills and general well-being, subjective meanings attached to gaming, and general comments on the usability of the devices and games. Staff observations and comments were incorporated in the data as a reflection base.

13.4.1 CASE #1: BALANCE BOARD GAMING

Two different groups participated in the tests. The first group consisted of people aged over 70 years old ($N = 34$). The average age of the participants was 86 years. The group was divided equally between men and women ($N = 17$). The subjective experiences and perceptions, including common free comments on the trial, were collected immediately after each player's experiment by a structured interview. The second group consisted of people with learning disabilities ($N = 23$). The age distribution of this group was slightly wider, as it varied from 21 to 60 years. As a result, the average age was 39 years. All the participants in the second target group worked in sheltered working facilities. The data were collected by theme interviews. The participants were asked to assess and describe their subjective perceptions of the games and the gaming event.

The experiences of participation in game playing were divided into two main categories: (1) positive experience and (2) negative or indifferent experience. Most of the participants in the first test group (88%) thought that playing was a positive experience. The participants described the experience as fun (32%), interesting (20%), easy to play (12%), very positive experience (9%), and challenging (9%). The game was also considered good for social activities and events (18%). Many people who did not play followed the event, and they were eagerly spurring those playing, which confirmed the social activation side of the gaming event.

The first test group gave in total four negative comments (12%). The negative comments were related to their personal conditions: "game controlling motion created pain," "my hands were shaking too much," "the game elements were hardly visible," and "just another toy to play with." The game was also experienced as rehabilitating and activating (94%). Only 6% of the participants were not sure whether the game was useful at all in activation or rehabilitation. The game was reckoned to assist in motor coordination (32%) and give sensible activation for arms and legs (24%) and activate brain (21%). People with impaired mobility thought that gaming was a rewarding and suitable rehabilitation method, whereas people who could participate in traditional physical exercises such as walks outdoors did not consider gaming as rewarding and suitable.

In general, the experiences of the game play were positive even among people with learning disabilities (the second test group). All participants (100%) found the game and the event entertaining, and most of them (87%) would play the game again at their homes. The feeling during the game play was described as comfortable (77%) or excited (23%). Excitement was related to the cat that appeared into the game and thus increased the difficulty level. In addition, excitement was caused because of the audience and because they experienced something completely new. Most participants focused only on the game play (91.3%), and the rest experienced that they were somewhat focused (8.7%). The gaming event was a very social and loud situation, where participants watching others play gave instructions and cheered to the others. All players (100%) also thought that the game provided leg exercise (the game was played by feet).

There were only a few negative comments in the second test group: "the loud audience was distracting" (7.7%), "I cannot imagine playing the game on my own" (13%), and "controlling the game with legs was weird" (4.3%). Some of the participants (17.4%) found the use of the "balance board controller" challenging, and 50% of the participants, who experienced controlling as challenging, were helped by the staff. One of the participants had ankle problems, but according to him, controlling became much easier when he had played for a while and got used to the equipment. There were also negative comments on slipping or getting the board in a wrong position on the floor.

The majority (74%) could play the game independently in the second test group, and the rest (26%) could play with some assistance from the staff or the researchers. The need for assistance depended on the player (verbal instructions, mental encouragement, or physical help). In each case, the need for assistance decreased after playing the game for a short time. The nature of game play was easily understood owing to well-known characters and their relationship (mice like cheese and dislike cats).

The participants of the trials related to Case #1, balance board gaming, were asked to give ideas for further game development and improvement. The ideas were partly utilized in the following cases. For example, the red squares and purple ball were changed to something else, based on the feedback.

In the elderly group, both the participants and the staff commented on the degree of difficulty of the game. Different difficulty levels were suggested by the elderly (9%) and the staff (30%), whereas some (12%) felt that a more challenging game would discourage participation. Some comments were directed at the coloring of the game elements, that is, they could not be easily distinguished.

People with learning disabilities thought that adding sound effects to the game (13%) or using different animal figures or animations would improve the game. Some comments were also made on the difficulty of the game, although the original game that was targeted at older adults had already been modified so as to adjust to the progress of the player (difficulty increased after getting a certain game score). The game ended for many players when the first cat appeared. Multiple chunks of cheese on the screen were also suggested in order to make scoring points easier.

Two participants discussed the size of the game elements. In their opinion, the elements should be bigger for visually impaired people or the game should be played on a wider screen.

13.4.2 Case #2: Tablet PC Gaming

In the 3-month test period, with the group of male war veterans ($N = 9$), we wanted to measure the possible effects of cognitively stimulating mobile games on older people's cognitive skills, when they played on a regular basis. The tests were arranged in an assisted living environment. The participants were in the age group of 88–97 years (average 90 year). They were diagnosed with mild or medium level of memory impairment but were cognitively active and lived in an assisted living environment.

User experiences and usability issues were collected after the test period by structured interviews. Each participant was interviewed privately in order to obtain as authentic and subjective information as possible. The focus was on their experiences of the trial period and the game's effects on their cognitive skills and general well-being. In addition, the users' previous experiences and use of mobile games and devices were explored. Staff observations and comments were incorporated into the data as a reflection base.

The results of the first game in Case #2 (tablet PC gaming) showed that the game was very engaging. Each player remained active during the whole 3-month test period. Before the 3-month gaming period, all participants participated in standardized mini–mental state examination (MMSE) and TMT-A tests. When the pre-trial and posttrial results were compared, only a couple of participants had improved their cognitive test results. Although gaming revealed no significant effect, further tests with much bigger test groups will be carried out in the future. However, the participants' experiences of the trial were outstandingly positive. Most of the participants thought that regular gaming was activating, interesting, and entertaining. In addition, they thought that the game provided suitable cognitive rehabilitation. Gaming helped them stay alert and active during the whole day. Otherwise, their days did not include too many cognitive impulses or things to do.

The second game in Case #2 (tablet PC gaming) followed the key idea of a standardized TMT A test, in which numbers are connected (by tapping, instead of drawing, the connecting line) in a correct order as quickly as possible. The TMT game had several difficulty levels. The participants were instructed to increase the difficulty level each week by adding five more digits on the screen. However, they did not increase the difficulty level when the game became too difficult. Instead, they continued playing the lower levels. The participants played the game rather successfully when there were 20 or less digits shown on the display. After the level with 20 or more digits, playing times decreased significantly. As seen in Table 13.1, only a few games were played at level 35.

The playing time of each completed level decreased over time, although the level became more difficult. This indicates that the elderly people learned and developed their skills in attention, reaction, and dexterity. Some players also decreased the error numbers and percentages over the trial period.

TABLE 13.1
TMT Game Shots per Difficulty Level

Difficulty Level	Played Game Shots in Total 729	Completed Shots	Success (%)
Level 5	23	12	52
Level 10	263	235	89
Level 15	167	161	96
Level 20	139	111	80
Level 25	62	44	71
Level 30	65	50	81
Level 35	5	3	60
Level 40	5	0	0

The majority of participants found the second game in the tablet PC gaming case easy to play, interesting, and suitable for cognitive activation. The usability of tablets improved outstandingly when silicon covers were used. They improved the grip of the tablet, especially for older people, because of their dry skin.

In general, tablets were considered easy to use. Apart from mental stimuli, the games offered useful activity to train finger dexterity and attention skills. The games also had an important role as a social activator, which involved the staff, visitors, and family members in competition and sharing experiences. The main problem detected was the poor functionality of the capacitive touch screen in case of the elderly, since their skin is dry. This naturally affected the playing time by increasing it. This sensitivity problem was not discovered in the developing process. When the problem was detected, we tried to solve it by a code to make the touch screen tapping as sensitive as possible. However, it did not help the users. This is one of the disadvantages of the otherwise effective capacitive touch screens, where function is based on the change in capacitance caused by an electrically conductive object such as a human touch. In future trials with older adults, a touch pen or touch gloves could be used to increase touch sensitivity. Another option is to employ resistive touch screens, which do not require any electrically conductive touching object, as it relays of the pressure.

13.4.3   CASE #3: BALANCE BOARD PAINTING

All the testers ($N = 17$) who tested the mobile painting game in the rehabilitation center of Kankaanpää reported that the playing situation had been a positive experience. The main finding was that people with more severe constraints enjoyed the painting more than other people. A possible explanation is that balance board painting enabled them to express themselves in a way that was not possible before. Some changes were also made to the mobile painting user interface on the basis of the feedback from the first tests (color picking, order of the elements, etc.). However, the graphical interface was not as important as the ability of the application to adapt to the user's skills in playing the game. Figure 13.6 shows how the users were able to use mobile painting, regardless of their state of disability.

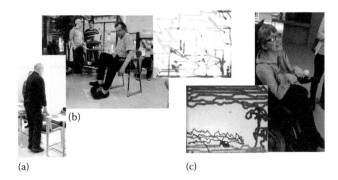

FIGURE 13.6 Balance board painting: (a) elderly people with a minor balance disorder painting while standing on the board, (b) people who are rehabilitated because of single-sided paralysis painting with the board as seated, and (c) person with significantly impaired mobility painting by using a light mobile phone.

Mobile painting was also tested with people with learning disabilities ($N = 12$) as a part of a summer event in a nursing home. Painting with the balance board was accepted well. All users said it was easy to use, despite the differences in their functional ability (some had a personal assistant). One specific goal was detected among all users. They all wanted to achieve something visible to show to others. The artwork itself was the main focus, and the exercise was left unnoticed. In conclusion, mobile painting was considered appealing. This was also one of our main goals when designing the application (to make the user experience the feeling of flow). This user group was also somewhat different in their cognitive abilities, but everybody could participate in the tests with the help of the staff and our testing personnel.

The third occasion where mobile painting was tested was a regional senior fair. Although elderly people were not as interested in the mobile painting app as people with reduced mobility or people with learning disabilities, everybody who participated in the test thought that the mobile painting app was interesting and rehabilitating. Using a mobile phone as a game controller was not smooth in the beginning, but the end result was always gratifying. They also felt that it was refreshing to get their own painting printed.

The tests showed that mobile painting was considered suitable for a group stimulating activity, because it promoted social interaction. Even people who did not participate in the test enjoyed mobile painting. Many people gathered around the testers and watched as participants did their work of art.

13.5   CONCLUSION

New technologies provide low cost and individualized and mindful ways to activate and motivate people to become self-supportive and vigilant. Especially in long-term care facilities, residents should be encouraged to maintain their cognitive, motor, and social skills by deploying new and easy-to-use technology to provide meaningful things to do both individually and in groups. Positive experiences make a difference in overall well-being. Easy-to-use technology is available, and there is growing evidence that both older people and people with special needs find this technology useful and also accept it. In the trials where the three different games were studied, the games were welcomed as a good tool for rehabilitation and recreation. The games give meaningful activities to those in care, thus saving the careers' time and effort. However, it is essential to make the games adjustable to personal skills and limitations.

REFERENCES

Alaska Department of Administration. (2013). Trail making test. http://doa.alaska.gov/dmv/akol/pdfs/UIowa_trail-Making.pdf. Retrieved July 6, 2013.

Csikszentmihalyi, M. (1975). *Beyond Boredom and Anxiety*. San Francisco, CA: Jossey-Bass.

Csikszentmihalyi, M. (1990). *Flow: The Psychology of Optimal Experience*. New York: Harper and Row.

Csikszentmihalyi, M., Abuhamdeh, S., & Nakamura, J. (2005). Flow. In A. Elliot (Ed.), *Handbook of Competence and Motivation* (pp. 598–698). New York: The Guilford Press.

Diaz-Bossini, J.-M., & Moreno, L. (2014). Accessibility to mobile interfaces for older people. In *5th International Conference on Software Development and Technologies for Enhancing Accessibility and Fighting Info-exclusion, DSAI 2013* (Vol. 27, pp. 57–66). Procedia Computer Science; Available at http://www.sciencedirect.com, Retrieved December 3, 2014.

Engeser, S., & Rheinberg, F. (2008). Flow, performance and moderators of challenge-skill balance. *Motivation and Emotion, 32*(3), 158–172.

Fairchild, J. K., & Scogin, F. R. (2010). Training to Enhance Adult Memory (TEAM): An investigation of the effectiveness of a memory training program with older adults. *Aging & Mental Health, 14*(3), 364–373. doi:10.1080/13607860903311733PMID:20425656.

Gao, Y., & Mandryk, R. L. (2012a). The cognitive benefits of playing a casual exergame. In *GRAND 2012*, Montréal, Québec, Canada.

Gao, Y., & Mandryk, R. L. (2012b). The Acute Cognitive Benefits of Casual Exergame Play. In *Paper on The ACM SIGCHI Conference on Human Factors in Computing Systems CHI 2012*, May 5–10, 2012, Austin, TX. Available at http://hci.usask.ca/uploads/256-p1863gao.pdf. Retrieved December 3, 2014.

Geda, Y., Roberts, R., Knopman, D., Christianson, T. J. H., Pankratz, V. S., Ivnik, R. J., ... Rocca, W. A. (2010). Physical exercise, aging, and mild cognitive impairment: A population-based study. *Archives of Neurology, 67*(1), 80–86. doi:10.1001/archneurol.2009.297PMID:20065133.

Khoo, E. T., & Cheok, A. D. (2006). Age invaders: Inter-generational mixed reality family game. *The International Journal of Virtual Reality, 5*(2), 45–50.

Kiili, K. (2005). Digital game-based learning: Towards an experiential gaming model. *The Internet and Higher Education, 8*(1), 13–24. doi:10.1016/j. iheduc.2004.12.001.

Kiili, K., de Freitas, S., Arnab, S., & Lainema, T. (2012). The design principles for flow experience in educational games. *Procedia Computer Science, 15*, 78–91.

Kiili, K., & Perttula, A. (2013). A design framework for educational exergames. In: de Freitas, S., Ott, M., Popescu, M.M. & Stanescu, I. (eds.), *New Pedagogical Approaches in Game Enhanced Learning: Curriculum Integration*, pp. 136–159. IGI GLOBAL, Hershey, PA.

Koivisto, A., Merilampi, S., Kiili, K., Sirkka, A., & Salli, J. (2013). Mobile activation games for rehabilitation and recreational activities—Exergames for the intellectually disabled and the older adults. *Journal of Public Health Frontier, 2*(3), 122–132. doi:10.5963/PHF0203001.

McCough, E. L., Kelly, V. E., Logsdon, R. G., McCurry, S. M., Cochrane, B. B., Engel, J. M., & Teri, L. (2011). Associations between physical performance and executive function in older adults with mild cognitive impairment: Gait speed and the timed "Up & Go" test. *Physical Therapy, 91*(8), 1198–1210. doi:10.2522/ptj.20100372PMID:21616934.

Merilampi, S., Sirkka, A., Leino, M., Koivisto, A., & Finn, E. (2014). Cognitive mobile games for memory impaired older adults. *Journal of Assistive Technologies, 8*(4), 207–223. doi:10.1108/JAT-12-2013-0033.

Sirkka, A., Merilampi, S., Koivisto, A., Leinonen, M., & Leino, M. (2012). User experiences of mobile controlled games for activation, rehabilitation and recreation of the elderly and physically impaired. In *pHealth Conference 2012*, June 26–28, Porto, Portugal.

Sirkka, A., Merilampi, S., & Leino, M. (2014). Mobiilipelit uudentyyppisenä kuntoutusmuotona muistihäiriöissä. Hyvinvointia edistävän teknologian tutkimusryhmä (HET). Satakunnan ammattikorkeakoulu, Sarja B, Raportit 4/2014. Pori, Finland. ISSN 2323- 8356 (online publication in Finnish).

Snowden, M., Steinman, L., Mochan, K., Grodstein, F., Prohaskam, T. R., Thurman, D. J., ... Anderson, L. A. (2011, April). Effect of exercise on cognitive performance in community-dwelling older adults: Review of intervention trials and recommendations for public health practice and research. *Journal of the American Geriatrics Society, 59*(4), 704–716. doi:10.1111/j.15325415.2011.03323.x PMID:21438861.

Spector, A., Thorgrimsen, L., Woods, B., Royan, L., Davies, S., Butterworth, M., & Orrell, M. (2003). Efficacy of an evidence-based cognitive stimulation therapy programme for people with dementia: Randomised controlled trial. *The British Journal of Psychiatry, 183*, 248–254. doi:10.1192/bjp.183.3.248PMID:12948999.

Swann, C., Keegan, R. J., Piggott, D., Crust, L. (2012). A systematic review of the experience, occurrence, and controllability of flow states in elite sport. *Psychology of Sport and Exercise, 13*(6), 807–819.

Tombaugh, T. N. (2004). Trail making test A and B: Normative data stratified by age and education. *Archives of Clinical Neuropsychology, 19*, 203–214.

Wilson, R. S., Deleon, C. F. M., Barnes, L. L., Schneider, J. S., Bienias, J. L., Evans, D. A., & Bennett, D. A. (2002). Participation in cognitively stimulating activities and risk of incident Alzheimer disease. *Journal of the American Medical Association, 287*(6), 742–748. doi:10.1001/jama.287.6.742PMID:11851541.

14 Home.com
Living in a Digital Environment

Charles G. Willems and Eric P.M. Hamers

CONTENTS

14.1 INTRODUCTION

What is living? Living is an active verb. It is something we all do each and every day, and each of us does it in a different way. Whether we live in the middle of a town, in a penthouse, in a cottage on the countryside, or in a college dorm, we live. Whether we are young or old, vital or not, venturous or subdued, we live. The way we live differs quite a lot. It is the result of an interaction between personal characteristics, environmental factors, and the activities we perform. We (try to) live in our own environment: our house. As a result of our own desires and possibilities, our houses are situated on different spots and differ in characteristics. The house chosen meets the resulting sum of our desires to change our house into a home. First of all, we want our home to be a safe and secure place to shelter, which we are able to share, which we choose to share, and to which we may give access. Our house is the starting point of our activities. This influences the filling of the space in the house. An international active athlete lives from a trunk that contains all the needed belongings, and a person raising children lives in a family house. Once we attempt to perform

255

professional activities in our house, decoration and facilities will change. The house is attributed to the level of usability and comfort we want to maintain. Finally, home decoration will show quite a lot of our identity—a sense of "this is I." The choices we make in the transition of a house into a home are motivated by our own identity and demonstrate to outsiders what we really care about [1].

The moment living and care provision (have to) go together, it will lead to changes in the house too. Especially to support people with special needs (PSN), who are in need of living support, a different arrangement of infrastructure in the house is required to optimize the delivery of living support adequately. It is this line of reasoning that enables us to discuss the opportunities given by domotics to support living, and from that, the opportunities of a so-called smart home will become clearer. Having discussed these issues, this chapter continues dealing with the potential of home technology in supporting care. Every function in care delivery that may become supported may give rise to the use of a new technology. This will also strengthen the need for an integrated platform. In the last paragraph, we will return to the user and describe the real challenge for applied research in assistive technology.

14.2   ENVIRONMENTAL SUPPORT

Living—as said—is making your home your place (see also Figure 14.1). Sometimes, assistance from the outside is needed to perform the living the way you really want it. Therefore, although one can imagine that living is considered an inside thing, it is the outside world as well that is contributing to men's well-being and men's health. During adult life, most time will be taken by living, working, and recreation. Shifting toward older ages, emphasis will be more on living

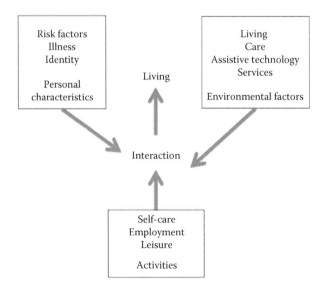

FIGURE 14.1   Living as a result of the interaction of three domains: personal characteristics, environmental conditions, and activities performed.

and recreation—indeed, a shift toward a smaller and more "inside home" circle. However, the environmental impact will be enormous as well. It is the outside environment that will make you feel living on the right spot on earth. During several years, people make connections to the physical environment of their house. Like a tree getting a solid roots work through the years, men, women, and families adapt their environment to move around, to get involved in social life, to feel safe, and to live! Therefore, attention has to be provided for this environment as well. Of course, not every street can be as described, but what should be aimed at is the possibility to connect people to their environment. Spatial planning is the keyword here. Connecting people to their environment seems obvious, but it should be taken into account that not every individual uses his environment equally. With increasing age, one's action radius decreases, thus making connection to nearby facilities more needed. Self-support and social cohesion within a smaller circle in this case become evidently more needed. Technology options in this context include digital platforms for small aides and social helpfulness. On the other hand, in cases where people endanger their personal safety by wandering around and become lost, a restriction may be needed to secure their safety. In that case, "track and trace" systems for the outside may be a useful support option for caregivers. For the inside, the use of technical devices to have certain doors closed may also be used to prevent dangerous situation. An overview of these kinds of monitoring technologies clarifies that technological developments allow for refinement in the functionalities involved and improvement of technical applications [2].

Our built environment can be considered a flexible business; depending on the time scaling, changes are minor or even drastic. Even if you remember the environment of your youth well, a visit after a prolonged period of absence may cause problems. Adapting to these changed situations may then be difficult. At the same time, a lot of technical evolutions will help in built environment to take care of difficult situations. Most common (already) and known is the navigation computer, making maps unnecessary. It will not be long until the elderly being confused or lost outside will be guided back to their safe haven via their hearing aid connection. For cars, this system is the state of art exactly. Difficulties with parking your car can be solved in the same way, by guiding you to the free spot of your wishes. Moreover, after having found your place, automotive technology can do your parking automatically.

Another concept where technology can have a future impact is the easy access to things, such as "Uber" for taxi services and "Airbnb" for hotel services. In future, these technologies can also help in making housing components available. On individual basis, this way it is possible to check in your own neighborhood guest-housing possibilities. Supply and demand makes a flexible built environment. Examples of movable assets largely exist, mentioning garden equipment or car and bicycle sharing. Accessibility, on the other hand, may also vary in relation to the digital world. Accessibility for the elderly is not likely to be exactly the same as for youngsters. Needs for different groups are different. Marketing campaigns have found its way to target groups; care-giving companies can do the same within the built environment. Ideas include providing help to the elderly or families, while providing music interaction to young people, thus targeting within the built environment as topic.

Spatial planning in built environment, as said, will be optimal when providing needs and whereabouts for the mix of inhabitants. In case of improper balance, certain groups may feel unheard. Therefore, accenting certain elements can contribute to a mix, which may influence community life to a higher standard. Shops for daily goods nearby and luxury goods off the center may contribute to this standard. Recreational emphasis for the elderly (such as strolling along greenery nearby) can mix with community-intensive get-together spots for families and youngsters. A golden formula for the winning environmental planning is hard to give, but consciousness is rising that its contribution to neighborhood's welfare is evident.

14.3 DOMOTICS AND LIVING SUPPORT

What do we mean once it is stated that a house has become a smart house? Research on the use of technology in living support has led to the concept of a "life-resistant house" [3]—a house that will meet the changing needs that exist during different phases of our life. The flexibility that is required to obtain this goal will be difficult to obtain once it is expressed in terms of space needed. In case the change can be met with a different use of the existing space, the transition will be much easier to accomplish. However, what are the consequences thereof to the technical specifications of the house? A research project performed with a consortium of housing and care organizations and technology suppliers has taken this question as a central issue. It led to the renewal of the concept of a "life-resistant house" and has been adopted by the housing and care organization "Humanitas" in Rotterdam [2]. Concerning the electrotechnical infrastructure, the conclusion at that time was that, in every room in the house, three infrastructures are needed: supply of electrical current (230 V), signal for cable TV, and connection for phone and data exchange. Starting from this central point in the room, every activity can be supported. At this point in time, almost 17 years later, we may note that TV and voice telecommunication may use the same infrastructure—wireless networks (WiFi, UMTS, 4G, etc). Yet, the essential point of 1999 result still holds. This knowledge has led to the development of a concept called the "residential gateway" [4], a technical module placed at a central spot in the house. In the Netherlands, this is mostly done near the entrance of the house. Starting from this point, the distinct infrastructures find their way through the house. At this spot also, the user interface is placed, by which the use conditions are determined. Technological developments continue. In 2001, a publication by Emile Aarts led to the concept of ambient intelligence [5]. It refers to the technology that is available "all over the place" and indeed, starting from the presentation of this concept, wireless networks and communication protocols for short (Bluetooth) and long ranges (Zigbee-UMTS) have become abundant in our life. As a result, we are constantly surrounded by this communication infrastructure. The vision by Aarts and others is gradually translated into products and applications that have entered the daily life of most of us, for instance, in activity support, comfort, organization, and delivery of new services. The European program for applied research "Ambient Assisted Living (AAL)" supports this development by supporting the cooperation between technologically oriented firms and user organizations [6].

14.4   APPLICATIONS TO SUPPORT LIVING

Taken from the perspective of a user, the use of technology in a house works best once the technology can be easily integrated in daily practice. As is often the case, each technology application uses its own technological means and protocols. Therefore, using technology as a supportive tool in living support more often leads to the use of a wide array of man-machine interfaces to control its function. In addition, in case of malfunction, a search regarding how to handle the control becomes a tedious task.

Modern technological developments more often use design and development principles that try to organize an intuitive use strategy. This can be accomplished once in-depth knowledge of the actual use is incorporated already during the design and development. User-orientated development or principles of co-creation are becoming regular practice in the development of applications, even for PSN [7,8].

During the development of home technology or domotics, this is also becoming regular practice. The ambition is to integrate informatics into applications for living support. In the ideal situation, this would mean that independent of the origin (= supplier, manufacturer) of the application, it can be used in connection to the already available infrastructure in the house, without having to organize adaptation. Although this appears to be simple, this represents a huge challenge. Consider, for instance, the use of electrical appliances. Modern housing in every European country uses electrical current. Yet, despite standardization activities at the European level, settings (voltage) and connections differ. Needless to say that in the information and communication technology (ICT) world, where there is supplier domination, the variation is even bigger. This continues once we have a look on the user interface operation and use instructions. The reasons behind this are well understood and of an economical nature. Choices made during past development activities have led to property rights and patents. Thus, changes in working processes may very well lead to a reduction in market value of earlier investments. Even worse, clients may retreat by no longer accepting updates and may start to use different brands to fulfill their needs. The internationally operating technology firms will try to safeguard their market position by joining forces with carefully selected competitors. The organization of a worldwide network, such as the "Continua Health Alliance," is an example of such a network, in which standardization aspects are discussed [9].

14.4.1   THE ORGANIZATION OF A NETWORK INFRASTRUCTURE

The organization of a so-called "user interface" is often a combined effort, in which disciplines work together. Domotics may organize a platform on which the integration of a technical support service can be done. The user interface plays an important role in obtaining the desired result. Requirements formulated by the user give input to the desired living conditions, for instance, temperature, lighting, and accessibility of the house. It should be warm when one enters the house or the light should go on when it is becoming dark and when someone enters the room. This is not only convenient or represents comfort, but this also contributes to the sustainable use of resources. Using this approach, it will become possible to organize certain tasks such as laundry during the lowest energy pricing. This support of sustainability is important to governmental ambitions as well as to consumers desires. It will become possible to control housing conditions from a distance by using mobile communication principles and smartphone technology [10].

14.4.2 ENVIRONMENTAL CONTROL

The desire to integrate to control various residential functions may originate from a need for support initiated by a decreased physical ability or because of a desire for comfort. Applications may include remote control of curtain movement, to be able to see who rang the doorbell, to control the climate in the home, and to have a variety of services organized. Usually, this leads to the fact that each technology comes with its own control. For users, this often represents an impossible situation. The need for systems integration is present. A transparent and a universal control structure becomes relevant. On the inside of the technical system, this leads to an infrastructure of several layers of communication protocols and "application programmable interfaces". On the outside, it must then lead to a simple and uniform access to the applications. On the inside, we often speak of "residential gateways." On the outside, we are talking about environment control or about the initiation of an alarm call. For users, young or old, it is always a problem to find the way to use such infrastructure. Evaluations of use by the intended users should be an important part of the R&D process to support the development and to obtain a user interface that works with a form of intuitive use [11].

14.5 HOUSING FOR THE ELDERLY AND THE USE OF HEALTH CARE SERVICES

14.5.1 ACTIVE ALARM

One of the oldest examples of a service that aims to support the elderly in living independently is the use of social alarms. In the Netherlands, its use started in the late 80s of the last century. The principle is very simple. An alarm device is installed in house; when operated by the push of a single button, a speaking-listening connection will be established with an employee of a care center. The availability of this service, 24 h a day for 7 days in the week, gives reassurance to the user that in the event of an emergency, the personal security situation can be supported appropriately. Research indicates that the use of social alarm is financed in different ways: by welfare services, by medical support, and by direct payment by the user. Especially for the elderly, this has become an important assistive technology to remain independent [12]. The development of social alarms systems for professional alerting is also used. In addition, the assistance is not offered from their own social network but by a collaboration of health care providers. In the delivery of support to the client who initiated a social alarm, a common problem is encountered; how to enable the care provider to enter the house of the initiator of the alarm? Initially, this led to the use of many key management systems. However, later on, these were integrated with the alarm system. This enables that access to the house is granted to the alarm successor as part of the emergency call procedure. The use of alarm infrastructure with an integrated door opening has become a support system for home care in the Netherlands and is financed under the homecare regulations. Yet there are still significant problems in use. Acceptance to use this technology by the user group that may benefit most is still a problem. Research from Sweden, based on practical experience with this technology of more than 40 years, has shown that this technology can also be limiting; this is because of the design that was used. The technology is especially designed to be used in a house-bound situation. This implies that the users, for their own sense of security, are obliged to stay at home. The social function would be enhanced if the alarm could be linked to a Global Positioning System (GPS), which would allow the users to keep a larger range in their mobility [13].

The technology for this is already available for some time, especially to support car mobility and finding the way. However, for a person who is going astray, timely and targeted detection by using GPS information is still weak. The positioning accuracy and the ease of use of the different tested systems still leave much to be desired [14]. It is clear that the need for such systems, for example, in the provision of care to people in incipient stages of dementia, is very large. It enables them to maintain a degree of freedom of movement and relieves their caregiver(s) in delivering support.

14.5.2   PASSIVE ALARM

The above-discussed active alarms have long been known. Technological developments allowed for the development of systems that do not require active participation of a user to generate an alarm. A so-called passive alarm is used to monitor activities. This involves the use of motion sensors that determine whether the movements are there in a room. Using an American system "QuietCare®" Proteion, a Dutch homecare organization, has built experiences. When sufficient information is available about the property and about the placement of the sensors and agreement is obtained on the interpretation of the messages, the behavior of the resident can be followed and interpreted at a distance. Central feature in the operation of the system is to detect changes in the behavioral pattern. This is achieved by comparing the activity pattern of the user with the record of the previous period. Examples of information gathered to build an activity pattern include data on the time of getting up, the total mobility, the number of meals actions, the occurrence of possible falls in the bathroom, and the sleeping pattern. When the observed behavior deviates from the previous data (limits are exceeded), the system automatically sends a warning signal to an equipped care center. Based on previously made arrangements with the customer, the information is interpreted and follow-up activities may be initiated, when needed. If agreed by the care receiver, an informal carer is also given access to the data, allowing him or her to render support, when needed. Technically, this system has natural limitations. The observations are disturbed when multiple people are present in the house, since all activities are summed up without indication to the person who performed the activity. Thus, the measurement system has value only if a single person stays in the house. Research shows that caregivers are supported by their activities. Moreover, the employees of homecare welcome the availability of this information; they gain insight into the activity pattern of their clients, which they could not get in any other way [15].

14.5.3 Medical Monitoring

Another form of the alarm uses the measurement of physiological parameters such as cardiac function and the insulin values. Depending on the parameter, it is possible to measure continuously by means of sensors (cardiac function) or through the intermediary actions of the client/user. In both cases, the data will be routed to a person present at a distance, who makes the interpretation of the transmitted values. This can be either real time or on delay. Depending on this interpretation, the client receives a feedback. In acute situations, a direct reaction can be initiated; in other situations, the client can be reassured, and this can go on with life. The success of this type of guidance depends highly on the way in which the chain of information is organized. Collaboration between professionals, who operate each by their own role, can be a great relief to the client. In this communication, technology plays an important role again. Research on supporting people with chronic heart failure has shown that the use of the "Health Buddy®" represents a simple and effective communication between patient staying at home and the accompanying specialist. The client answers a set of questions, automatically generated, on a daily basis. In this way, the professional is able to monitor the situation and give guidance, when needed. Meanwhile, a growing number of health care management programs (heart failure, diabetes, chronic obstructive pulmonary disease, asthma, and hypertension) are used this communication [16].

The client can also use this technology to support the development of healthy behavior pattern. The applications discussed above are often housebound applications. The development of mobile communications has led to the availability of portable systems. The development of blood pressure (BP) application by MobiHealth may serve as an example. This company has developed the BP@Home web application, by which the client can himself measure BP at any place and share info with a therapist [17]. Supported by crowd funding, developments continue leading to applications that support monitoring of life functions at any time and any place. See, for instance, recent developments by the Silicon Valley firm, Scanadu [18].

Internationally, there are many projects in development to support people using targeted physiological information. It may be expected that as the knowledge about the use of such systems will increase, the preventive value will be clearer. From the perspective of knowledge dissemination, it will be necessary to pay attention to the way in which this information is shared with stakeholders. After all, as we have already seen in the use of social alarms, pre-eminently a safety device, not everyone is inclined to use them. Especially with regard to adaptation to the use of this kind of support systems and equipment and to choosing mechanisms to deliver individual feedback, a substantial improvement in our knowledge position is required.

14.5.4 The Use of ZorgTV

Especially in the case of housing the elderly, the TV is extensively used as a mode of communication to the outside world. It is an excellent medium to see what is happening in the world and to support the private relaxation. A disadvantage of this medium is that until now, only one-dimensional communication was possible. With the advent of digital communication, use of two-way communication is booming. The ability to access a wide bandwidth enables the support of video communication. The first participants using video communication by two-way cable TV were found in Kortrijk via "tele senior" project in 2003. Moreover, in the Netherlands, examinations usability of this application was initiated rapidly. However, despite the efforts of all parties involved, the pretty hefty investments that were associated with it have long proved to be unsuccessful. Research evaluations demonstrated potential benefits [19]. Others had the opportunity to look with a more care-related manner into the possibilities of video communication in healthcare. FocusCura, a technology-oriented firm, in conjunction with a consortium of Health Maintenance Organizations (HMOs), was able to develop a number of applications of CareTV.

In the Netherlands, a nationally funded transition program for long-term care played an important role in the development of applications of video communication in healthcare in a way that the business is supported positive. The development of social business cases based on the practical experiences thereby constituted the vehicle. The research that has been done shows that the participants feel supported and that both their emotional and social loneliness can be reduced [20]. The number of users is steadily increasing. A critical success factor to the development of such a facility is the ability to develop new care services that can be supported by this infrastructure. In recent times, the use of the iPad (a handy platform using WiFi to support communication) has become very prominent in longitudinal care in the Netherlands.

14.6 FROM INDIVIDUAL SYSTEMS TO AN INTEGRATED PLATFORM

From the components described above, it is not hard to imagine that this development still has no end. After all, technical developments continue, and frequently, they create new opportunities. Yet, it also shows that it is difficult to organize sustainable developments. Each new application, having been supported with a new technical system, cannot be good. Personally, we are also convinced that it will be necessary to join forces. Organizations are not able to adapt to changing infrastructures that are needed to support new functions. It is not sustainable to restart again and again with a new technological infrastructure. There are other developments in the area of care and well-being that introduce changes. In Figure 14.2, a schematic representation of these developments are given. The message is clear: the use may vary but the communication infrastructure is as similar as possible. For the care organization Proteion, the perspective outlined initiated reasons to engage with service providers and local authorities in order to align the goals themselves. Concentrated power in communication support delivers more efficiency in its corporate activities. This observation is also made by other organizations.

When we take a look at the developments in the mobile communications, another perspective is easier to imagine. Steve Jobs initiated a revolution that gave the Apple company a solid position in mobile communication. The analysis of the characteristics that have contributed to this success reveals that it is not just a clever marketing strategy. Is that the sole reason that a lot of people are crazy enough to spend the night on the street in order to get the latest gadget from Apple? Not really! A clear vision on the usefulness of the products is also a key success factor. Without an

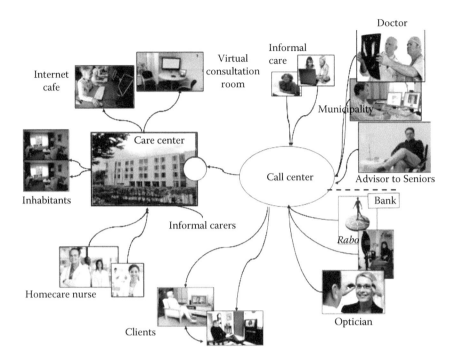

FIGURE 14.2 Schematic presentation of a virtual platform to support interactions at the community level.

extensive study of a manual, everyone eventually find the way with the equipment. The design results in a clear approach on how to use different applications (apps) to obtain the desired utility. Each app has an easily recognizable logo. Each user can choose from different apps to get done what is wanted. Whether it is obtaining information from the rain radar, separate phone numbers, the stock market page, your page on "Facebook" or "LinkedIn", or even to control the thermostat in the living room, it can be done from this device and at a distance. This is the power of the approach initiated by Apple. The nice thing is that not only this development may be useful in the individual's own phone, but you can give it a place in a "civic community." The baker, the butcher, the insurance adviser, the neighborhood association, home care, the newspaper, and the town, they all have their own apps. Once all the residents of a neighborhood are connected in this way, social network is created. With this in mind, the company Aristo has developed the social application "Cubigo" [21]. Utilizing the principles of co-creation, an infrastructure is originated that may connect closely with the individual needs of a participant.

The thought "Who has the youth, has the future" will undoubtedly guide this development further. Notwithstanding the fact that older people are also very well able to go along with these developments. In 2013, at least 50% of the group aged between 65 years and 75 years use the Internet on a daily basis for e-mail and Internet banking [22]. In other words, even the group of older persons goes along with this development. The Internet has become an important means for seeking and finding health-related information. In research on the navigation behavior of younger and older people on the Internet, eye-tracking technology was used to measure how the information-seeking behavior manifests itself. It has been found that the daily use of the Internet has more impact on our navigation skills than the age of the user. In short, even if there is no Internet experience accumulated during the formative years, it is a deficiency that can be compensated for by using it [23]. In other words, if data on the activities in the area are available on a virtual community platform, the (older) local residents who attach importance to that information will be stimulated to participate in a platform. By using technology in this way, a modern meaning may be given to social networks. With the help of this method, it is then possible to specify a modern meaning to social networks—in a way that caters to their own interest, their own needs, and their own opportunities to be socially active.

We want to achieve this level of use of new support, although it is obvious that this will not happen automatically. It will also not work once it is endorsed solely by a technology-oriented vision. No, it should be clear that the power of this development must be sought in the participation of all concerned. In short, it represents a challenge for applied research.

14.7   CONCLUSIONS

In this article, we have attempted to paint a picture of the perspective given by living in a digital world. In the support of both living and well-being, as well as in supporting care, many new possibilities will arise, even related to the support for and use by the elderly persons. To demonstrate in an irrefutable way that the addition of this technology to everyday activities allows us to live longer in independence is difficult, if not impossible. This is, in part, a dilemma, since many of these technological developments are initiated by or supported financially from that societal perspective. This requires a focused argument of sense and nonsense. For aged persons, it is as good as impossible to deny these new developments. Yet, once they get connected to the benefits of these new applications, they will be stimulated to exploit them. Once the principles of co-creation are more and more used during the development stage, it will become easier to identify the benefits of these applications. The concept of ambient intelligence offers many, yet unsuspected, possibilities to support us in living. Living in a digital world will, if you want to recognize it, remain an exciting journey, which will also provide opportunities for PSN. The involvement of the users in the design and implementation process will continue to be an essential condition for the success. This is truly a challenge for applied researchers.

REFERENCES

1. Willems, C.G. (2012). Domotica en slim wonen. In: E. Wouters, J.J. van Hoof (ed.), *Zorgdomotica* (pp. 73–79). Utrecht, the Netherlands: Lemma Utrecht.
2. Peetoom, K., Lexis, M.A.S., Joore, M.A., Dirksen, C.D., De Witte, L. (2014, September). Literature review on monitoring technologies and their outcomes in independently living elderly people. *Disability and Rehabilitation Assistive Technology*. doi:10.3109/17483107.2014.961179.

3. Willems, C.G., Lieshout, G., van Dijcks, B.P.I, van Parijs, M.M. (1999). The comfort project. In: C. Buhler, H. Knops (ed.), *Assistive Technology on the Threshold of the New Millennium* (pp. 513–518). Amsterdam, the Netherlands: IOS Press.

4. Kester, J. (2005). Zelfstandig blijven met Domotica. Report published by ECN Petten.

5. Aarts, E.H.L., Harwig, H., Schuurmans, M. (2001). Ambient intelligence. In: J. Denning (ed.), *The Invisible Future* (pp. 235–250). New York: McGraw-Hill.

6. Active and Assisted Living Programme-Europe. http://www.aal-europe.eu webpage visited on December 22, 2015.

7. Sixsmith, A.J., Gibson, G., Orpwood, R.D., Torrington, J.M. (2007). Developing a technology 'wish-list' to enhance the quality of life of people with dementia. *Gerontechnology*, 6(1):2–19.

8. Span, M., Hettinga, M., Vernooij-Dassen, M., Eefsting, J., Smits, C. (2013). Involving people with dementia in the development of supportive IT applications: A systematic review. *Ageing Research Reviews*, 12:535–551.

9. Continua Alliance. http://www.continuaalliance.org webpage visited on December 22, 2015.

10. Honeywell Inc. http://www.kijkvoelbeleef.nl: "evohome" webpage visited on December 22, 2015.

11. Cremers, G., Spreeuwenberg, M., van der Heide, L., Spierts, N. (2010). *Malberg; Evaluatie van menustructuren.* Heerlen, the Netherlands: Rapport Zuyd Hogeschool, lectoraat Technologie in de zorg.

12. Hogens, A., Jansen, B.B., Pouw, N., Rous, J. (2003). Personenalarmering in Nederland. Amsterdam Rigo research en AdviesBV rapport 83100.

13. Boström, M., Kjellström, S., Malmberg, B., Björklund, A. (2011). Personal emergency response system (PERS) alarms may induce insecurity feelings. *Gerontechnology*, 10(3):140–145.

14. van der Leeuw, J., van der Heide, F., Willems, C. (2009). Zoeken en gevonden worden?! De inzet van GPS voor mensen met dementie Eindrapport bruikbaarheidsonderzoek. Utrecht Vilans.

15. Willems, C.G., Spreeuwenberg, M.D., van der Heide, L.A., De Witte, L.P., Rietman, J. (2011). The introduction of activity monitoring as part of care delivery to independently living seniors. In: L. Bos, A. Dumay, L. Goldschmidt, G. Verhenneman, K. Yogesan (eds.), *Handbook of Digital Homecare 3: Successes and Failures (Communications in Medical Care and Care Compunetics)* (pp. 167–180). Heidelberg, Germany: Springer Verlag.

16. Sananet. http://www.sananet.nl webpage visited on December 22, 2015.

17. Mobihealth. http://www.bpathome.nl visited on December 22, 2015.

18. Scanadu. http://bit.ly/1NJ2o4t visited on December 22, 2015.

19. Boonstra, A., Broekhuis, M., van Offenbeek, M., Westerman, W., Wijngaard, J., Wortmann, H. (2008). *Kijken op afstand een leerzaam alternatief+ Rapport.* Groningen, the Netherlands: RUG.

20. Willems, C.G., Spreeuwenberg, M.D., van der Heide, L.A., De Witte, L., Rietman, J. (2011). Care TV to support care delivery to independently living seniors. In: L. Bos, A. Dumay (eds.), *Handbook of Digital Homecare 3: Successes and Failures (Communications in Medical Care and Care)* (pp. 31–42). Heidelberg, Germany: Springer Verlag.

21. Cubigo. http://www.cubigo.com visited on December 22, 2015.

22. Dutch Central Bureau of Statistics (CBS). http://www.cbs.nl/nl-NL/menu/themas/vrije-tijd-cultuur/nieuws/default.htm internet gebruik ouderen fors toegenomen visited on December 22, 2015.

23. Loos, E.F. (2010). Het gebruik van oude en nieuwe media: de (ir)relevantie van leeftijd. *Geron*, 12(20):13–17.

Index

Note: Page numbers followed by f and t refer to figures and tables, respectively.